DAZZLED AND DECEIVED

also by Peter Forbes

The Gecko's Foot: How Scientists are Taking a Leaf from Nature's Book

DAZZLED AND DECEIVED
MIMICRY AND CAMOUFLAGE

PETER FORBES

YALE UNIVERSITY PRESS
NEW HAVEN AND LONDON

WITHDRAWN

For information about this and other Yale University Press publications, please contact:
U.S. Office: sales.press@yale.edu   www.yalebooks.com
Europe Office: sales@yaleup.co.uk   www.yaleup.co.uk

Set in Arno Pro by IDSUK (DataConnection) Ltd
Printed in Great Britain by TJ International Ltd, Padstow, Cornwall

Library of Congress Cataloging-in-Publication Data

Forbes, Peter, 1947–
    Dazzled and deceived: mimicry and camouflage / Peter Forbes.
      p. cm.
    Includes bibliographical references and index.
    ISBN 978-0-300-12539-9 (alk. paper)
    1. Mimicry (Biology) 2. Camouflage (Biology) I. Title.
    QH546.F66 2009
    578.4'7–dc22
                                                                        2009023577
A catalogue record for this book is available from the British Library.

10 9 8 7 6 5 4 3 2 1

For Diana

CONTENTS

ILLUSTRATIONS

Plates

Figures

ACKNOWLEDGEMENTS

Drawing on stories from natural history, art and warfare, this book owes a great deal to many disparate sources. I am grateful to the biologists and others who agreed to be interviewed and/or answered my queries: Sean Carroll, Lars Chittka, Mariella Herberstein, Blanca Huertas, Chris Jiggins, Mauricio Linares, Leena Lindström, the late Michael Majerus, Steve Malcolm, James Mallett, Antonia Monteiro, Michael Nachman, Fred Nijhout, Antony Penrose, David Pfennig, Bob Reed, Richard Stokes, John Turner and Dick Vane-Wright.

Others helped with specific enquiries: Toshiharu Akino, Stephen Badsey, Wilhelm Barthlott, William Connor, Malcolm Edmunds, Thomas and Maria Eisner, Wittko Francke, Helen Ghiradella, Clare Goddard, Wendell Haag, Marcus Kronforst, Alistair Maskelyne, Florian Schiestl, Peter Skelton and Dieter Zimmer. In some cases I was not able to use fascinating material because of the demands of space and the narrative thread.

Especial thanks are due to Christopher Potter, who read the entire manuscript and made countless helpful suggestions; to John Turner, who shared his immense knowledge of mimicry, read some of the chapters and even managed to smuggle some poetry into our correspondence; to Chris Jiggins, who read the text and kept me *au fait* with his latest research findings; and to Mauricio Linares, who shared his quest for the Missing Link. Ann Charlton at the University Museum of Zoology, Cambridge, was especially helpful in my researches into the work of Hugh Cott.

When told about my project, friends, colleagues and family came up with interesting suggestions and helpful leads: my thanks to Anthea Arnold, Julian Bell, Mark Divall, Neil Forbes, Philip Kreager, Colin Marr, Catherine Maxwell, Rafael O'Dwyer, Ruthie Petrie, Leo Reich and Mike Sharpe. John Rushton at STV kindly provided a copy of *Magic at War*.

Dr Johnson said that 'a man will turn over half a library to make one book', and in this case the library was the British Library, with its astonishingly rich science collection. It would be churlish not to acknowledge also the importance of Google in making available the latest research and in retrieving references or quotations that have gone astray – Google is so often a lifeline. It is hard now to imagine research without it.

The London Butterfly Centre, now sadly defunct, was a source of inspiration. Hopefully, the projected Butterfly House near St Albans, to be housed in an enormous butterfly Eden-Centre type dome, will make good this lack. In the USA, we already have Butterfly World: www.butterflyworld.com.

My tireless agent Andrew Lownie made it possible for a decades-old dream to be realised and Heather McCallum, my editor at Yale University Press, played a powerful role in shaping the book.

Above all, I have to thank my wife, Diana Reich, who read at least two full drafts, kept her nerve and suggested drastic reorganisation at crucial stages when the material threatened to topple out of control.

For the use of copyright material and illustrations and kind permission to refer to manuscripts, rare editions and archives, my most grateful acknowledgements are due to the British Library, London; the British Library Newspapers, Colindale, London; the National Archives, Kew, London; the Imperial War Museum, London; the Bodleian Library, Oxford; the Cambridge University Library; the Library of the Natural History Museum, London; the University Museum of Zoology, Cambridge; Archive Services, the University of Glasgow; the Royal Academy Library, London; Selwyn College Archives, Cambridge; the Scottish National Gallery of Modern Art; the Lee Miller Archives, Chiddingly, East Sussex. Crown copyright material at the National Archives at Kew, London, and at the Imperial War Museum, London, is reproduced by kind permission of the Controller of Her Majesty's Stationery Office.

For permission to quote from published works and unpublished papers, acknowledgements are due to the following: to Farrar Straus and Giroux LLC for excerpts from III and IV of 'The butterfly', from *A Part of Speech*, by Joseph Brodsky. Translation copyright ©1980 by Farrar Straus and Giroux LLC. Reprinted by permission of Farrar, Straus and Giroux LLC; to Curtis Brown Ltd, London, on behalf of the Estate of Winston Churchill, for an extract from a wartime speech, © Winston S. Churchill; to Mrs Ruth Lyons for permission to quote from the papers of Hugh Bamford Cott; to P.J. Placito for permission to quote from the papers of E.B. Ford held in the Department of Special Collections, Bodleian Library, Oxford; to Belinda Farley for permission to reproduce extracts from Sir Alister Hardy, *The Living Stream* (Collins, 1965); to Glasgow University Archives Services for permission to quote from the papers of Sir John Graham Kerr; to Phaidon Press Ltd for permission to quote an extract from *The Image and the Eye* by E. H. Gombrich, first published in volume form 1982, reprinted 1986, 1994, 1999, 2002 by Phaidon Press Limited © 1982 Phaidon Press Limited, www.phaidon.com; to Random House, Inc for an extract from Primo Levi, *The Periodic Table* (Schocken, 1984); to Richard Mabey for an extract from Richard Mabey, *In a Green Shade* (Basil Blackwell, 1983); to Alistair Maskelyne for extracts from Jasper Maskelyne, *Magic – Top Secret* (Stanley Paul, 1949); to the Rt Hon The Viscount Montgomery of Alamein CMG CBE for permission to quote from the papers of Field Marshal Montgomery in the Imperial War Museum; to the Wylie Agency for permission to reproduce extracts from Vladimir Nabokov: *The Gift* © 1963, Dimitri Nabokov; *Speak Memory* © 1947, 1948, 1950, 1951, 1967, Vladimir Nabokov. All rights reserved. By permission of the Estate of Vladimir Nabokov; to Rozsika Parker for permission to quote from the published and unpublished writings of Dame Miriam Rothschild; to Lady Scott for permission to quote from the papers of Sir Peter Scott MBE held in the Cambridge University Library; to Bloodaxe Books for permission to reproduce an extract from 'The spirit is such a blunt instrument' by Anne Stevenson, from *Poems 1955–2005* (Bloodaxe Books, 2005); to the Trustees of the Imperial War Museum and to Sergeant Bob Thwaites for permission to quote from his papers in the Department of Documents, Imperial War Museum; to A. P. Watt Ltd on behalf of the Literary Executors of the Estate of H.G. Wells for permission to reproduce an extract from 'The moth', from *Collected Short Stories* by H. G. Wells; to George White for permission to reproduce an extract from Nelson C. White, *Abbott H. Thayer* (Connecticut Printers, 1951); to Camilla Wilkinson for permission to reproduce extracts from Norman Wilkinson, *A Brush with Life* (Seeley, 1969).

All attempts at tracing the copyright holder of Geoffrey Barkas, *The Camouflage Story* (1952), Cassell plc, a division of the Orion Publishing Group, were unsuccessful. Every effort has been made to trace the copyright holders of the extracts included in this book. The author and publishers apologise if any material has been included without permission or without the appropriate acknowledgement, and would be glad to be told of anyone who has not been consulted.

I am grateful for an Award from the Authors' Foundation of the Society of Authors.

PROLOGUE

As I write, a comma butterfly has just flown into the house, intending to hibernate. It sits motionless in the window, with its wings closed. Because it has to survive the winter without moving, it is a brilliant leaf mimic: the wings are scalloped ornately as if the margins had been nibbled, totally disrupting the typical butterfly wing pattern. Waves of mottled grey to brown spread across the wing, just like the patterns on a dead leaf, and there is a tiny crescent shape in white on each wing, like a fleck of bird's droppings. It is a tattered leaf, which can lie unnoticed throughout the winter.

The beauty and purpose of the comma's leaf mimicry are obvious. Since vision first developed in creatures in the Cambrian era, around 540 million years ago, appearances have been hugely significant in shaping the course of evolution. The art historian Sir Ernst Gombrich (1909–2001) was interested in more than paintings: he took the whole field of visual representation as his province and he wrote beautifully on mimicry and what nature's copying and stylised warnings mean for the art of human beings: 'For the evolution of convincing images was indeed anticipated by nature long before human minds could conceive this trick ... the art historian and the critic could do worse than ponder these miracles. They will make him pause before he pronounces too glibly on the relativity of standards that make for likeness and recognition.'

Gombrich finds various styles of art in nature: a leaf butterfly can fancifully be considered to be 'a naturalistic artist', natural selection having produced a facsimile of the dead-leaf pattern. But the eyespots sported by some butterflies are stylised gestures: 'They represent, if you like, the Expressionist style of nature.' Then there are patterns whose meaning is established by association. Creatures which are protected by toxicity or powerful defences such as stings signal their danger with bright warning colours: in temperate regions, wasps and ladybirds are typical examples. The warning colours of toxic species – red, yellow, black and white – are also the colours we use for our road signs and for hazardous materials logos. This may be an innate response of living creatures to bright colours, which stand out against the duller greens, browns and greys of vegetation and rock. Perhaps the colours have become associated with danger over long periods of evolutionary time, and so have only a conventional meaning, in the same way in which the names of objects are assigned by convention (a road is not intrinsically a 'road', but also a 'rue', an 'ulitsa' or thousands of other names in different languages). At any rate, the parallels between natural and human warning coloration are too obvious to ignore.

Deception has always played a large role in human affairs. Early humans – hunter gatherers – were as adept in the wiles of camouflage as any animal: they had to be.* The biblical story of Esau and Jacob has an uncannily biological feel to it. When the 'smooth man' Jacob wanted to impersonate the 'hairy man', his hunter brother Esau, in front of their blind and aged father Isaac, he covered himself with hairy goatskins. Isaac recognised Jacob's voice but preferred to trust in the smell and hairy feel of the goat hunter – smell and touch being more primal and elemental senses than hearing: 'The voice is the voice of Jacob, but the hands are the hands of Esau.'

Ambiguity and disguise exert a powerful hold on the human imagination. Mistaken identity, whether intentional or accidental, has permeated

---

* Strictly, the word 'camouflage' did not exist before it was coined by the French in 1917, during the First World War. The word derives from the verb *camoufler*: to dress up, disguise. Before its use in war, it had theatrical connotations and the terms 'concealing coloration' or 'protective coloration' were used instead. 'Camouflage' is so much the better word that I use it throughout, even when it is technically anachronistic. 'Protective coloration' can also mean, not concealment, but display.

myths, legends and literature. In particular, in the Oedipus legend or in many of Shakespeare's plots – *As You Like It, The Merchant of Venice, Measure for Measure, A Midsummer Night's Dream, Twelfth Night, The Winter's Tale* – falsifying a person's identity is the engine of many human dramas. In human society now, these doubts are put to the court of DNA testing; and in the realm of nature DNA reveals that many species are not what they appear to be on the surface.

Most deceptions in the human realm have a parallel in nature: when the Trojans led the Horse into their city, unsuspecting that it was packed with Greek soldiers, they were at one with the ants who, fooled by deceptive chemical odours, carry caterpillars of some blue butterflies into their nest and feed them in preference to their own larvae. The passage in *Macbeth* in which the men of Malcolm's army disguise themselves as trees in their approach to Dunsinane prefigures the modern sniper, decked out in tattered rags, twigs and leaves.

Until the end of the nineteenth century, troops wore intimidatory, bright uniforms, red being a favourite – as it is among the many toxic species which proclaim their dangerousness by conspicuous display. But in the later years of the nineteenth century the increased power of weaponry, especially the machine gun, made concealment a priority.

Camouflage in war has been a test case of the 'two cultures'. In normal times, attempts to find links between art and science are well meaning but mostly of no avail, because the territories and methods of their disciplines seem so different. But in camouflage in wartime, both biologists and artists felt they had unique expertise to offer. Now the territory was the same, but the mindsets were very different. The story of the attempt to apply lessons from nature's camouflage and mimicry to military combat in the two world wars has many intriguing twists, involving three principal naturalists – Sir John Graham Kerr, Hugh Cott and Sir Peter Scott – besides a cast of artists, soldiers and one magician. The convergence of nature, art and science in the world wars occupies the central period covered here: between the explorations of the pioneers and the hi-tech genetic work of today. As a recent themed issue of the Royal Society's *Transactions* recognises: 'Camouflage research has for a significant length of time linked biology, art and the military, stemming from the work and influence of Abbott Thayer and Hugh Cott.'

The artistic aspect of mimicry and camouflage lends an added dimension to the allure of the subject. I began to collect examples of mimicry twenty-five years before beginning this book, at a time when I was working as a natural history desk editor. At that stage I saw mimicry primarily in terms of descriptive natural history, with strong aesthetic overtones. The stories in Hugh Cott's *Adaptive Coloration in Animals* (1940), which was at the time still the most compendious and authoritative text on the subject, enthralled me with their revelations of surprising mimicry in exotic settings.

But mimicry is more than a fantastical tale of visual punning in nature: it can tell us so much about evolution. The story of evolution is the biggest prize in biology. To understand how living things develop from the egg; how the genes act to shape the organs; to know in depth how inheritance works; to unravel the genetic basis of disease and to devise cures: these are the normal goals of most biological research. But the more we know about organic processes in the here and now, the more we learn about the three-and-a-half-billion-year journey from the first replicating molecules to the current, multi-million species cornucopia of life. One of the most startling findings of molecular biology has been the revelation that all creatures are closer cousins under the skin than we had imagined. The question 'Are you a man or a mouse?' can now be answered 'Both' – since around 99 per cent of human genes have a mouse equivalent. It seems that large differences in the form and function of animals have been achieved by relatively minor genetic mutations.

But evolution did not occur only by shuffling a pack of genetic cards. Whether genes can be passed on at all depends on how an organism lives in a real environment with other creatures. Understanding how genes and the total environment interact over vast stretches of time is an enormously complicated issue and we don't have video footage of the past, but to look at the evolution of a restricted phenomenon can be a test case. Mimicry and camouflage – the resemblance that one life-form has either to another or to a part of the environment – have some special features which make them ideal for studying evolution in action. And the best subjects are insects – spiders, mantises, but above all butterflies and moths. There are many reasons for this. The rapid turnover of generations in insects and their vulnerability mean that natural selection bears very heavily on them. And butterflies are special among insects because their whole being is displayed on their wings – two-dimensional objects are easier to study than anything with three.

Very often we fall into circular arguments when we speculate on evolution because there is no purpose to it – no end in sight. We see that some organisms have survived, so we say that these must have been the fittest. And which are 'the fittest?' Those which have survived. But in mimicry one creature has led and another has, through selection, copied it. The old problem of attributing to nature a goal when it has none dissolves in the face of mimicry because, although there is no purpose to the whole of it, for the mimicking species there *is* a goal: to copy the model. So we have an index of the success of evolution in producing the match. Similarly with the butterflies that mimic dead leaves. Success is demonstrable.

So, if we discovered the pattern-forming genes in a mimic and its related but non-mimetic cousins, we would learn something about adaptation, natural selection and the genetic mechanisms behind a particular instance of precise copying. Obviously camouflage is somewhat different, but no less intriguing. When a moth exquisitely disappears against the bark of a tree, it tells us something about the passive I-am-a-camera nature of the genetic pattern-forming processes; but there will be no correlation between the genes which make the bark in a tree and the genes which make the patterns on the moth's wings. In mimicry, we will see how some quite similar pattern-making genetic machinery has been tweaked in one butterfly to match another.

As the genes behind mimetic resemblances and the time scale on which mimicry was achieved are revealed, the course of evolution will be reconstructed. Evolution leaves traces of its innovations in the DNA of present-day organisms: not a simple record that we can read as a timeline of development, but rather as forensic clues to be matched against fossil evidence. The story of evolution will probably be revealed first in mimetic and camouflaged creatures. I cannot think of a more enticing project, involving beautiful creatures in luxuriantly populated environments and the shifts in DNA which lie behind them.

CHAPTER 1

DARWINIANS, MOCKERS AND MIMICS

Some of these resemblances are perfectly staggering – to me they are a source of constant wonder & thrilling delight. It seems to me as though I obtain a glimpse of an intelligent motive pervading nature, as well as of the mighty, never-resting wonder-working laws that regulate all things.

Henry Walter Bates, Letter to Darwin, 1861

Mr Bates had come a long way. For eleven years – 1848–59 – this son of a Leicester hosier had journeyed through the Brazilian Amazon: from Para (now known as Belém), 70 miles from the sea, to St Paulo, about 1,800 miles inland (Fig. 1.1). He had been robbed of his money, suffered from chronic ill health and sometimes had acute bouts of fever. He had lived by sending home, via an agent, specimens of the wildlife he encountered. Despite his hardships, he revelled in it. This is what he had dreamt of when, as a youth, he collected butterflies and beetles in Charnwood Forest, Leicestershire. When he arrived in Brazil he was twenty-three.

Everything here in the Amazon was superabundant and larger than life. Bates's first scientific paper, published just before he left England, was called 'Notes on Coleopterous insects frequenting damp places', a title redolent of drab, dour England. At Para there were 700 identifiable species of butterflies just within an hour's walk of his *rocinha*, the country-house he had made his base, when in the whole of the British Isles at the

Fig. 1.1 Bates' and Wallace's travels through the Amazon: 1848-1852 (Wallace) and 1848-1859 (Bates). Bates reached St Paulo, within 200 km of the Peruvian border; Wallace explored the more northerly Rio Negro and Rio Uaupes. The key staging posts have all changed their names since the travellers were there: Para (now Belém), Barra (Manaus) and Ega (Tefé).

time no more than sixty-six species of butterfly were thought to exist.* But what butterflies there were, here in the Amazon:

---

* The present tally is only 59 species of historically resident butterflies, with three more regular migrants that breed (including the red admiral and painted lady). More than 30 other species have been found in Britain as vagrants or accidental introductions. So thorough were the Victorian naturalists that no new butterflies were found in Britain in the twentieth century.

Amongst the lower trees and bushes numerous kinds of *Heliconii*, a group of butterflies peculiar to tropical America, having long narrow wings, were very abundant. The prevailing ground colour of the wings of these insects is a deep black, and on this are depicted spots and streaks of crimson, white, and bright yellow, in different patterns according to the species. Their elegant shape, showy colours, and slow, sailing mode of flight, make them very attractive objects, and their numbers are so great that they form quite a feature in the physiognomy of the forest, compensating for the scarcity of flowers.

Henry Walter Bates (1825–92) was not just a collector of these gorgeous insects. He had set off with his friend Wallace, later to become famous as Darwin's co-discoverer of evolution by natural selection, with the purpose of 'solving the problem of the origin of species'. Bates was not disappointed. Not only were there hundreds of species of butterflies, but, when their wing patterns and other features were compared, many curious anomalies emerged.

Bates was a sturdy, vigorous man with a rather swarthy, incised face and penetrating eyes. He was a prime example of the largely self-educated Victorian. The eldest of four brothers, at the age of thirteen Bates was apprenticed to one of the city's characteristic hosiers, Alderman Gregory. He was not obviously destined for any kind of fame. But Bates had great curiosity and educated himself at the Leicester Mechanics Institute. There he studied Latin and Greek, French (languages were a forte), drawing and essay writing. His favorite book (which he was once again re-reading on his death-bed) was Gibbons' *Decline and Fall of the Roman Empire*. He also learned to play the guitar. But his greatest passion was nature: beetles – which are the largest single group in creation – were his first love, then butterflies.

When Bates was twenty, his master hosier died and his apprenticeship ended. He was forced to take up clerical work and was unhappily employed at Allsop's brewery in Burton upon Trent when he realised that his destiny lay far away from the malt-laden air of dreary, red-bricked streets in the English Midlands. In 1844, in a serendipitous encounter, Bates had met another passionate naturalist, Alfred Russel Wallace (1823–1913), and they would soon be plotting a journey unimaginable to most. Wallace in his memoirs could not recall where the two had met, but Bates believed it was in the Leicester Public Library.

Wallace was, like Bates, a self-made man. After his father's death he was forced to leave grammar school at the age of fourteen to earn his living, and he set himself up as a surveyor in Bedfordshire with his older brother William. There he began to appreciate wildlife and, through his surveying, geology. Like Bates, Wallace attended a mechanics institute* and by the age of twenty-one he was sufficiently competent to begin teaching at the Collegiate School, Leicester.

In 1846 Wallace left Leicester for his native Neath in South Wales following the death of his brother, but he kept in touch with Bates, who came to visit in the summer of 1847 and to hunt beetles. Both had begun to speculate on the variability and origin of species. They read *Vestiges of the Natural History of Creation* (1844) by Robert Chambers, founder of the well-known *Chambers' Encyclopaedia* and an amateur naturalist. *Vestiges* was an erratic text which nevertheless encouraged many Victorian naturalists to ask how species were formed. Bates and Wallace began to think that the workings of nature must be more apparent in places where they were also at their most various and abundant. And that meant the tropics. But where exactly?

Both young men had read, and were inspired by, Darwin's account of his voyage on the *Beagle*: 'as the journal of a scientific traveller, it is second only to Humboldt's "Personal narrative" ', claimed Wallace. But the prime catalyst was another book: W. H. Edwards' *A Voyage up the Amazon*, published in 1847. This was a highly coloured, stereotyped portrait of exotic jungle scenery by a brash American explorer which made an impression on the young would-be adventurers. The prose is a kind of naturalist's pulp non-fiction: 'At this period too, vast numbers of trees add their tribute of beauty, and the flower-domed forest from its many altars sends ever heavenward worshipful incense . . . Monkeys are frolicking through festooned bowers . . . Squirrels scamper in ecstasy from limb to limb . . . coatis are gamboling among fallen leaves . . .' The trees seem to have sprung into being 'at this period', and no animal lacks

---

* Beginning around the time when Bates and Wallace were born, mechanics institutes spread to most Victorian towns. Often founded by philanthropist manufacturers, they were an important source of adult education for working men.

its nursery verb of motion. Happily, when the two came to write up their travels, they cast off this garish influence.

Although Bates was still only twenty-three and Wallace was two years older, they quickly agreed to embark on an expedition to Brazil, which they would fund by sending back their specimens to London. The keeper of butterflies at the British Museum (the natural history section of which officially became the Natural History Museum in 1990) had assured them that northern Brazil was a happy hunting ground.

Bates and Wallace left Liverpool on 26 April 1848, on a small trading vessel, and they arrived at Para, Brazil on 28 May. Para was a town of about 15,000 inhabitants which Bates found to be attractive but decayed. He admired the white Italianate buildings with their red-tiled roofs and the streetscape punctuated by palms. But Para's recent history had been turbulent. Briefly, from 1808–22, Brazil had been the centre of the Portuguese Empire – which was a consequence of the Napoleonic Wars in Europe. It then seceded from Portugal, becoming an independent empire but still dominated by the Portuguese and under a Portuguese 'emperor'. Its highly complex racial mix of indigenous Indians, Portuguese and black Africans (some of whom were slaves), and much intermarriage, resulted in a wide range of recognised mixed races.* This society was volatile, and a violent slaves' revolt against the Portuguese ruling class in 1835 had reduced Para's population by 40 per cent.

But the town was surprisingly benign when Bates and Wallace were there, only thirteen years later. Bates comments on the many Catholic festivals, some lasting as long as nine days, and on the harmonious atmosphere at these times:

> When the festival happens on moonlit nights, the whole scene is very striking to a newcomer. Around the square are groups of tall palm trees, and beyond it, over the illuminated houses, appear the thick groves of mangoes near the suburban avenues, from which comes the perpetual ringing din of insect life. The soft tropical moonlight lends a wonderful charm to the whole.

* *Mameluco* = white and Indian; *mulatto* = white and black; *cafuzo* = Indian and black; *curiboco* = *cafuzo* and Indian; *xibaro* = *cafuzo* and black.

In Bates's account of the journey, expectations of the exotic rain forest constantly come up against dawning reality. The region is rich, but not in every way. Deep rain forest is so dominated by trees that every living form has to adapt or perish. Flowers are few, because the trees crowd out the light. Few flowers mean few of the insects which feed on their nectar and pollinate them, and no bees. The beetles, so dear to both explorers, were not to be found on the ground because ferocious large ants dominated that niche. Mammals were few and mostly of the tree-dwelling kind such as the monkeys.

But profusion and drama there were, when you knew where to find them. Gradually, the Amazon yielded up its treasures. The trees were a source of the most constant wonder: the major forest canopy had sheer trunks, with no branches below one hundred feet. The life of the forest crown, where many birds lived, really was another realm. The trees were so dominant and the climate created by them, with abundant rainfall, so moist, that many secondary trees, known as epiphytes, lived without roots, by twining around the stems of the principal trees and reaching for the light.

From the first days the two men encountered the giant Sauba leaf-cutting ants, which were as dangerous as rats, invading houses and stealing food. To stop them, Bates resorted to blowing up the ant trails with gunpowder. Then there were the six-inch hairy spiders, the boa constrictor snakes, and the Massaranduba or cow tree, which produces a copious supply of a milk-like drink from its bark, as pleasant as cow's milk (left to stand, it thickens and then can be used a as glue, to stick broken pots).

The Amazon challenged the perceptions of men from the northern hemisphere. The climate is more or less even throughout the year, so there are no seasons: animals and plants go through staggered cycles of reproduction and growth instead of the synchronised waves of seasonal climates. Para was not as hot as many other places near the Equator, rarely rising above 34°C. But it was humid. On landing, Bates observed: 'The hot moist mouldy air, which seemed to strike from the ground and walls, reminded me of the atmosphere of tropical stoves at Kew.'

Bates and Wallace would rent a house for a while and then explore on foot. Deeper excursions such as their first, a three-month trip up the Tocantin River, involved hiring a boat and crew and again renting a house as a base at their destination. After six months they realised that it would be much more valuable to split up. The Amazon rain forest is so large that it made sense for the two to cover different territories.

Bates journeyed through the jungle, along the course of the Amazon and some of its tributaries. The forest canopy goes on for ever, apparently all the same. But Bates found that the butterflies inhabiting different regions of the forest varied enormously. More striking still was that some butterflies with very similar appearances seemed, on anatomical grounds, to belong to entirely different species:

> In the shady ravines of the forest, many species of Ithomiae were found in greater or less abundance . . . Flying amongst the Ithomiae was now and then to be observed a *Leptalis* . . . they are very interesting because they have come to imitate, each a species of *Ithomia* . . . In fact I was quite unable to distinguish them on the wing; and always on capturing what I took for an *Ithomia*, and found when in the net to be a *Leptalis* mimicking it, I could scarcely restrain an exclamation of surprise.

From the names alone, this might not sound so remarkable, but the Ithomiae belong to the dazzling family Heliconidae, the typical butterflies of the rain forest, with black wings splashed with crimson, yellow and white patterns. The *Leptalides*, which were flaunting similar colours, belong to the family of whites: the Pieridae. They were cousins of our familiar cabbage whites. Somehow the *Leptalides* had acquired red and yellow pigmentation, and in just the right places.

The naïve naturalist (which Bates was not) has no doubt that the way a creature looks is what it is. And as far as butterflies go the wing patterns are all. The bodies of butterflies have many minute distinguishing points, but only the expert with a microscope can discern most of them. Since the great eighteenth-century Swedish naturalist Carl Linnaeus (1707–78) began the ordering of creation into a pattern of relationships, much attention had been paid to these minute anatomical features, particularly the genitalia. Nevertheless, at the first sight of a butterfly, a tentative identification comes from the wing patterns. This, Bates found, could be misleading.

Bates had discovered that one butterfly had somehow borrowed the external appearance of another. He called the phenomenon mimicry, the word we still use today. Colloquially, he sometimes called the copycat species 'mockers'. We know mimicry as a human trait. We are always amused when toddlers begin to repeat back what we say, usually with touching inaccuracy, and anyone who can replicate the gestures of someone famous will always be the life and soul of any party. So to find this mimicking trait in 'blind' nature was shocking and thrilling.

Another word about those names: *Ithomia* and *Leptalis*. They are Latin of course. Most people prefer their species to have homely names: for butterflies, names such as the blue plumbago-cloudless sulphur butterfly or the silver washed fritillary have the right kind of ring. But, in encountering hundreds of species in the Amazon, such names were not to hand. And Bates loved these Latin names. In the fly-leaf of his Latin school grammar he had written:

I am as fond of Latin
As women are of satin.

Latin does not change, being dead, but, unfortunately, many of the butterflies Bates observed have suffered name changes since he wrote, which adds an element of confusion to the mystery of mimicry.

Not only did *Leptalis* mimic *Ithomia*, but the *Ithomia* changed its pattern from region to region, and the *Leptalis* which copied it also changed, in order to maintain the mimicry. Bates found many examples of these copycat shadowings during his travels. And it was not just a case of shadow *pairs* of species. In some places, a group of half a dozen or more different species all showed the same wing patterns. Some were not even butterflies at all, but day-flying moths. These were deep mysteries, which would take some time to unravel.

Bates and Wallace were reunited for two months at Barra in early 1850, somewhat less enthusiastically than at their first meeting, to judge from Wallace's account. It rained incessantly and Wallace recalled:

I now had a dull time of it in Barra . . . Mr Bates had reached Barra a few weeks after me, and was now here, unwilling, like myself, to go further up the country in such uninviting weather . . . Between two and three months passed away in this unexciting monotony.

They then split up again to explore different tributaries of the Amazon, Wallace going north on the Rio Negro and Bates south on the Solimoens River. They didn't meet again till both were back in England ten years later.

In 1852 Wallace was forced by ill health to return to England. It took him almost three months to sail home and he suffered shipwreck by fire (with the loss of all his specimens and notes), ten days exposed to the elements before being picked up, and severe storms in a rotting boat – 'one of the slowest old ships ever' – unfit for such voyages. His first ship, the *Helen,* was carrying an inflammable cargo of both familiar and exotic tropical products: rubber, cocoa, annatto, balsam-capivi, and piassaba.* He is remarkably dispassionate in his account of the disaster, taking a forensic interest in why the ship had caught fire and learning that balsam-capivi tends to combust spontaneously. It needs to be packed in small kegs cushioned with damp cloths, but a last-minute consignment was packed with inflammable rice chaff. Then, when the ship was filling with smoke, instead of sealing up the smouldering compartment, holes were opened up with axes, letting in the air. The result could have been predicted: 'It now presented a magnificent and awful sight as it rolled over, looking like a huge caldron of fire, the whole cargo of rubber, etc., forming a liquid burning mass at the bottom.'

Arriving in London on 1 October 1852, Wallace at first vowed never to travel again, but he was soon making plans. Even before the disastrous homecoming, Wallace's Amazon journey, recounted in *A Narrative of Travels on the Amazon and Rio Negro,* was less successful than Bates's. Wallace's most important contribution would come after he had returned to England and indeed completed other arduous journeys.

Back in England, Wallace began to reflect on the variable species he and Bates had observed. Unknown to him, someone else was pondering

---

* Annato is a red colouring and flavouring agent used in many cheeses. It now has an E number (E160b). Balsam-capivi (also known as copaiba) is an oily resin used in varnishes and ointments. Piassaba is a strong, coarse fibre derived from palm trees and used to make ropes, brushes and brooms.

the same questions: Charles Darwin (1809–82). Darwin's long journey towards his theory of evolution by natural selection is one of the great personal odysseys.

Unlike Bates and Wallace, Darwin came from an educated family with a remarkable intellectual record. His grandfather, Erasmus Darwin (1731–1802), was one of the most illustrious figures of the English Enlightenment: a polymath with highly progressive views and an inventive streak. He believed in a primitive form of evolution and wrote about it, in his voluminous book *Zoonomia* (1796): 'shall we conjecture, that one and the same kind of filament is and has been the cause of all organic life'. Erasmus is also credited with the first mentioning of the notional Big Bang which began the Universe. Charles acknowledged the influence of Erasmus's views on his own developing thought.

Erasmus Darwin had been a member of the Lunar Society of Birmingham, a high-minded dining club that included several of the most prominent scientists, industrialists and artists in Britain, the makers of the Industrial Revolution. They included James Watt, inventor and manufacturer of the first commercial steam engines; the chemist and co-discoverer of oxygen Joseph Priestley; Josiah Wedgwood, the founder of the pottery dynasty; and Joseph Wright, who was the court painter to the society.

Beginning as a young man of no especial intellectual attainment and with no sense of vocation, Darwin had been the despair of his physician father; he spent his youth shooting wildlife and, while at Cambridge, enjoying the company of 'some dissipated low-minded young men'. Before Cambridge, two years at Edinburgh studying medicine had failed because Darwin could not stomach the gore.

The young Darwin was the type that, a century later, would be known as 'hearty'. His father admonished him: 'You care for nothing but shooting, dogs, and rat-catching, and you will be a disgrace to yourself and all your family.' Darwin's saving grace was that his passionate interest in nature went beyond shooting it, pleasurable though that was. From an early age he had the collecting bug, and what he most coveted were indeed bugs – beetles. He recounts to what lengths he went to acquire them:

One day, on tearing off some old bark, I saw two rare beetles, and seized one in each hand; then I saw a third and new kind, which I could not bear to lose, so that I popped the one which I held in my right hand into my mouth. Alas! it ejected some intensely acrid fluid, which burnt my tongue so that I was forced to spit the beetle out, which was lost, as was the third one.

The young Darwin might have neglected his formal studies, but he had a knack of charming various interesting senior figures into pouring their considerable knowledge into his mind. At Cambridge he was introduced to John Henslow (1796–1861), the Professor of Botany, and never looked back. The two were so often seen together on botanical expeditions that Darwin became known as 'the man who walks with Henslow'. He then met Adam Sedgwick (1785–1873), the noted geologist. Darwin was field assistant to Sedgwick on a geological excursion in North Wales in 1831, and this crash course in geology proved invaluable for his future work.

At Cambridge Darwin was preparing for life as a country parson. He was diverted from this destiny (in which he would no doubt have amassed a tidy collection of beetles and other creatures) by the opportunity, presented by Henslow in 1831, to join HMS *Beagle* as the scientific observer for a voyage to South America whose official purpose was to complete the mapping of the coast of this continent for the Admiralty. Given how important this voyage was to become, it is remarkable in retrospect by what slender threads Darwin's participation hung. His father raised grave objections and was only mollified by Darwin's uncle, Josiah Wedgwood II, son of the founder of the Wedgwood pottery company, Josiah Wedgwood (1730–95).

Darwin was imperfectly equipped for the task in more ways than one. On the voyage he suffered terribly from sea sickness, and he admitted on return to England that 'from not being able to draw and from not having sufficient anatomical knowledge, a great pile of MS. which I made during the voyage has proved almost useless'. Yet despite his amateurism Darwin observed wonderful things during his long voyage around South America (1831–6). When he began, he was still to some extent the young man who liked to shoot and had not yet become a fully fledged naturalist. He reflected on this fact:

> Looking backwards, I can now perceive how my love for science gradually preponderated over every other taste. During the first two years my old passion for shooting survived in nearly full force, and I shot myself all the birds and animals for my collection; but gradually I gave up my gun more and more, and finally altogether, to my servant, as shooting interfered with my work, more especially with making out the geological structure of a country.

Geology provided his first revelations. Armed with Lyell's *Principles of Geology* (1830) and with his own field work with Sedgwick, Darwin felt able to interpret the rock formations he had seen. Sir Charles Lyell (1797–1875) taught that the processes of geology had been more or less constant throughout the history of the earth and that rocks had been thrust up from beneath the sea for huge distances. Darwin encountered just such a formation, 6,000 feet high, in the Chilean Andes. The rocks contained the fossil remains of trees similar to living *Araucarias*, which had grown at ocean level. Here, high in the Andes, there were no living trees – all was barren.

In the Pampas of Argentina he found fossil bones of many extinct animals, the most dramatic being those of the giant *Megatherium*, a ground sloth which Darwin recognised as a relative of the living tree sloths, but very much larger. At this time, the idea that many species had become extinct was as controversial as the idea that they had evolved. And, unlike the living sloths, which have given their name to idle behaviour, *Megatherium* was judged to be capable of 'undermining and hawling [*sic*] down the largest member of a tropical forest' – something no living animal can do. Why would differently sized versions of the same basic design inhabit the earth at different times?

And then Darwin came to the Galapagos Islands – volcanic formations, some of recent eruption, with very strange animals. Not only were many of the latter unique to the islands, but Darwin discovered – too late to investigate the phenomenon properly – that each of the ten or so islands had its own distinctive version of each species:

It has been mentioned, that the inhabitants can distinguish the tortoises, according to the islands whence they are brought. I was also informed that many of the islands possess trees and plants which do not occur on the others . . . Unfortunately, I was not aware of these facts until my collection was nearly completed: it never occurred to me, that the production of islands only a few miles apart, and placed under the same physical conditions, would be dissimilar. I therefore did not attempt to take a series of specimens from the separate islands. It is the fate of every voyager, when he has just discovered what object in any place is more particularly worthy of his attention, to be hurried from it.

Darwin was not thinking in evolutionary terms during his voyage; he merely observed and recorded diligently. He brought back vast collections of specimens. It was only when he brooded on his findings, back in England, that he began to wonder at the degree of variation in the species he had seen.

We think of Darwin as the Galapagos voyager, but most of his productive life was spent at his home, Down House, at the foot of the North Downs in Kent. Here, from 1842 on, he settled to wrestle with his big idea. From 1837 he had kept a notebook in which he recorded examples of variation and the possibility that one species might be 'transmuted' (the word he used at the time) into another.

In 1838 Darwin read Malthus's *An Essay on the Principle of Population* and almost immediately conceived the idea of natural selection. Thomas Malthus (1766–1834) was a highly educated English country parson from a well-to-do family who shocked the world by discussing human populations and the disparity that exists between the rate at which people breed and the provision of food. He believed that the population would always grow to outstrip the food supply, the imbalance only being rectified by catastrophes delivered by one or more of the Four Horsemen. Darwin generalised this to all creation – animals will tend to reproduce at a rate above the generational replacement rate. As he later formulated it in *On the Origin of Species*:

A struggle for existence inevitably follows from the high rate at which all organic beings tend to increase. Every being, which during its natural lifetime produces several eggs or seeds, must suffer destruction during some period of its life, or during some season or occasional year, otherwise on the principle of geometrical increase, its numbers would quickly become so inordinately great that no country could support the product. Hence, as more individuals are produced than can possibly survive, there must in every case be a struggle for existence, either one individual with another of the same species, or with individuals of distinct species, or with the physical conditions of life.

Darwin realised that, if species generated variation when they reproduced naturally (which he had observed), some variations would inevitably be better fitted to survive in the struggle against starvation and predation. It was not possible that all variations would be equally successful. In *On the Origin of Species*, Darwin would come to write eloquently of the struggle that must ensue:

When we look at the plants and bushes clothing an entangled bank, we are tempted to attribute their proportional numbers and kinds to what we call chance. But how false a view is this! Every one has heard that when an American forest is cut down, a very different vegetation springs up; but it has been observed that the trees now growing on the ancient Indian mounds, in the Southern United States, display the same beautiful diversity and proportion of kind as in the surrounding virgin forests. What a struggle between the several kinds of trees must have gone on during long centuries, each annually scattering its seeds by the thousand; what war between insect and insect – between insects, snails, and other animals with birds and beasts of prey – all striving to increase, and all feeding on each other or on the trees or their seeds and seedlings, or on the other plants which first clothed the ground and thus checked the growth of the trees!

In coming to his conclusions, Darwin had investigated domestic selection very profoundly, spending much time in the company of pigeon fanciers, dog and cattle breeders and the like. A large part of the *Origin* is concerned with variation, both in nature and in domestic animals. Darwin discovered that domestic breeders could achieve huge changes in the form of an animal in the course of a single human lifetime, simply by observing very small changes in the offspring and constantly selecting for a particular trait over many generations, until it was sufficiently amplified to produce a major change.

A very powerful argument in Darwin is the fact that although these breeders had expert and intimate knowledge of the changes they had wrought, within a few human lifetimes the changes were so profound that the chain that led back to the wild species was lost from collective memory:

> Ask, as I have asked a celebrated raiser of Hereford cattle, whether his cattle might not have descended from longhorns, and he will laugh you to scorn. I have never met a pigeon, or poultry, or duck, or rabbit fancier, who was not fully convinced that each main breed was descended from a distinct species.

But Darwin believed, and modern DNA analysis has confirmed, that the domestic races of animals are descended from a single common or garden wild species in each case.

Darwin reasoned that selection – in creating such different animals as the dachshund and the German shepherd dog by means of only very slight changes at each breeding – was so powerful that not even these master breeders, after a passage of time, could believe that all these forms stemmed originally from a wild animal, probably only a few hundred or at most a few thousand years before. He then reasoned that, if man could wreak such changes in a very brief time by selecting for only small natural variations, the struggle for life, also acting on small variations but over immense periods of time, must be so much more powerful: 'How fleeting are the wishes and efforts of man! How short his time! and consequently how poor his products will be compared with those accumulated by nature during whole geological periods.' On these grounds, Darwin felt sure that natural selection acted on no more than the usual small variations seen all the time in wild populations, but over immense periods of time. This powerful idea has exerted a huge influence ever since, and has generated very powerful polemics for and against it. The most serious difficulty was the need for each evolutionary step to confer some advantage: surely the intermediate stages between one well adapted creature and another would be good for nothing?

Darwin's leading critic in this was to be St George Jackson Mivart (1827–1900). Mivart made the argument for the uselessness of what he called 'incipient stages' very forcefully and this troubled Darwin greatly. The problem could hardly be resolved with the knowledge of the time, but Darwin was able to satisfy himself that Mivart was wrong to suggest that a half-evolved eye would be useless. Many creatures have a low degree of light sensitivity and this is certainly useful to them. In the early, dark days of evolution, in the kingdom of the blind, the weakly light-sensitive organism would have been king.

Darwin's answer to objections such as Mivart's was to plough on, questioning nature, amassing as much information as he could. His voyage had left him chronically indisposed and he would not travel again. Instead, at Down House, he undertook a vast correspondence with naturalists, plant and animal breeders. And he was a typical Victorian paterfamilias. The personal and professional life of distinguished people often throws up ironic contrasts. As a biologist, Darwin came to believe in the benefits of out-breeding for producing healthy variations; as a man, he belonged to a somewhat inbred elite.

Charles Darwin's father, Erasmus's son, had married a Wedgwood – Susannah. Like father, like son: on 29 January 1839, Charles Darwin married his cousin, Emma Wedgwood. Darwin, possessor of Wedgwood genes, married a Wedgwood. The question of inbreeding would not have occurred to him when he got married, but, eventually, after the studies that led him to publish *The Effects of Cross and Self-Fertilization in the Vegetable Kingdom* (1876), the fact that some of his children had illnesses similar to his own suggested to him that marrying his cousin might have contributed to his children's ailments.

Darwin's crisis came in the spring of 1851, when his eldest and favourite daughter, Annie, died at the age of ten from a 'Bilious Fever with typhoid character'. Darwin's doubts about the consequences of marrying his cousin, the doubts about religion that his work had engendered, and the personal sense that the loss of this wholly good and still immature girl could not be reconciled with the notion of any kind of moral deity fused into a knot of misery that never quite left him. Philosophically, from this point on, when the intensity of his sorrow had receded, Darwin maintained a cool, level stance in the face of the claims of creationists, saying that he did not see why a beneficent God would proudly introduce to the world such beings as the ichneumon wasps, which grow as larvae inside the bodies of caterpillars, the eggs having been injected there by the female. Many other grotesque life cycles could be cited.

To complete the picture of evolution, Darwin needed to understand inheritance. Favourable traits spread through the population by a means whose mechanism was totally unknown. Darwin conducted many plant-breeding experiments in his glasshouses and garden in an attempt to understand this vital aspect of evolution. The mechanism of inheritance remained elusive, but he *was* able to show that those plants that were cross-pollinated by insects were stronger and more vigorous than those which had been self-fertilised. One hundred and fifty years later, with a wealth of genetic knowledge at our fingertips, we know very well why this should be so: inbreeding allows harmful recessive genes to accumulate in a population, weakening it.

For twenty years after Darwin conceived the idea of natural selection in 1838, he tried to buttress his theory with more evidence. He was very reluctant to publish and only slowly divulged the idea to anyone. In 1847 he showed the botanist Joseph Hooker (1817–1911) a 231-page manuscript which he referred to as 'The Big Species Book', for Hooker to read and comment on. Hooker suggested that, before he pronounced on the origin of species, Darwin ought to know a little more about *existing* species. Darwin threw himself into the study of barnacles, publishing a two-volume monograph on the subject in 1854.

Meanwhile Wallace was travelling again, now in the Malay Archipelago, looking for evidence on the formation of species. In 1855 he published an early theory of evolution, 'On the law which has regulated the introduction of new species', but it did not specifically broach the idea of natural selection. Darwin read this paper but apparently did not appreciate the risk it posed that his idea would be pre-empted. His friend Sir Charles Lyell did: in May 1856, Lyell visited Darwin and urged him to publish immediately to ensure his priority. Darwin vacillated.

Following Wallace's 1855 paper, Darwin and Wallace began a correspondence, although some of their letters have been lost. Darwin reported that he was well advanced on 'the Big Species Book', and they discussed the evidence for evolution openly; but Darwin never referred to the concept of natural selection.

Matters came to a head in 1858. By now, Darwin's 231-page manuscript had expanded to ten chapters and 250,000 words – it was indeed 'The Big Species Book.' In February of that year, in Sumatra, a bed-ridden, feverish Wallace made the same deduction from Malthus that Darwin had made in 1838. He wrote up his theory in a single night. On 9 March 1858, Wallace sent his paper to Darwin. It took till June 18 to reach him, but this time Darwin got the point. He wrote to Lyell:

> Your words have come true with a vengeance that I should be fore-stalled. You said this when I explained to you here very briefly my views of 'Natural Selection' depending on the Struggle for existence. – I never saw a more striking coincidence. If Wallace had my M.S. sketch written out in 1842 he could not have made a better short abstract! Even his terms now stand as Heads of my Chapters.

For a while Darwin thought that his project had been ruined by Wallace's paper. He agonised in a letter to Lyell: 'I would far rather burn my whole book than that he or any man should think that I had behaved in a paltry spirit.'

What followed was one of the most surprisingly amicable and honourable deals in intellectual history. Lyell and Hooker encouraged the two men to present a joint paper to the Linnean Society, which was and is a leading forum for natural history, founded in 1788 in honour of the founder of modern zoological nomenclature, Carl Linnaeus. The historic presentation of the Darwin/Wallace paper took place on 1 July 1858, in the company of some thirty members of the society. Neither Darwin nor Wallace attended. Lyell and Hooker were the paper's sponsors, but no more converts were made that evening. The earth did not shake. Presumably many of the traditionalists present hoped that this dangerous new idea would fizzle out. The president of the society, Thomas Bell, later made the notorious remark that the year's meetings had not 'been marked by any of those striking discoveries which at once revolutionize, so to speak, [our] department of science'.

The paper was published on 20 August 1858, and Darwin now set himself to finish his book. In fact the urgency of the situation spurred him to abandon 'The Big Species Book' and to begin again, constructing a simplified narrative – not a scientific monograph but a lively polemical text that anyone might read. *On the Origin of Species* was published on 24 November 1859, and it sold out on the first day. Predictably, conservative and clerical interests attacked the book instantly. The set-piece confrontation between the irreconcilable forces occurred in Oxford at a meeting of the British Association for the Advancement of Science on 30 June 1860. Darwin's public champion was Thomas Henry Huxley* and the opposition came from the Bishop of Oxford, Samuel Wilberforce.† Wilberforce made the mistake of resorting to the cheap debating ploy of sarcasm. He asked whether it was on Huxley's grandfather's or grandmother's side that he was descended from an ape. Accounts differ as to what Huxley actually said but the gist of it was: 'If the question is put to me, "would I rather have a miserable ape for a grandfather than a man highly endowed by nature and possessed of great means and influence and yet who employs these faculties and that influence for the mere purpose of introducing ridicule into a grave scientific discussion" – I unhesitatingly affirm my preference for the ape.'

In the popular mind, the struggle between evolution and religion has always been seen as a battle for the story of human origins: did we evolve from lower creatures – ultimately from a few self-replicating chemicals about 3.5 billion years ago – or were we divinely created? But Darwin's *On the Origin of Species* very carefully refused to discuss the origins of humans: the book was about all the *other* species.

The launch of the idea of natural selection galvanised naturalists everywhere and none more so than Bates, who had returned to England only months before the *Origin*'s publication. He remained in the Amazon for seven years after Wallace had left, but eventually illness wore him down too and he was forced to return home. But this was good

---

* Thomas Henry Huxley (1825–95) was known as 'Darwin's Bulldog' for his strong advocacy of natural selection. He was the founder of an intellectual dynasty, which included the biologist Julian Huxley and the novelist Aldous.
† Samuel Wilberforce (1805–73) was the third son of William Wilberforce, the prime mover in the abolition of slavery. His rhetorical skills earned him the nickname 'Soapy Sam'.

timing, for the researches of Bates and Darwin were highly complementary. Darwin was hungry for examples of natural selection in action and Bates needed a theory to explain what he had seen in the butterflies of the Amazon: there was a convergence of the twain. Darwin and Bates began to correspond in September 1860: about variation, species . . . and mimicry.

Bates was not the kind of naturalist content merely to collect species and to engage in formal discussions as to where they could be placed in the Linnaean scheme. He wrote to Darwin: 'As to ordinary Entomologists they cannot be considered scientific men but must be ranked with collectors of postage stamps & crockery.' Here he anticipated the great twentieth-century physicist Rutherford, who widened the scope of the 'stamp-collectors' jibe to include all scientists except physicists. Bates, like Rutherford, wanted to understand nature's *processes*. He wanted to understand what lay behind the mimicking butterflies he had seen in the Amazon rain forest.

Bates was aware that there was something special about the Heliconidae,* the large family of longwing butterflies dressed in bright postal reds, yellows, blues and tiger-stripe orange patterns. If *Leptalis* butterflies, members of the cabbage white family, got themselves up to look like Heliconidae, this was obviously a Darwinian adaptation. There was some advantage in their looking like this:

> It is not difficult to divine the meaning or final cause of these analogies. When we see a species of Moth which frequents flowers in the daytime wearing the appearance of a Wasp, we feel compelled to infer that the imitation is intended to protect the otherwise defenceless insect by deceiving insectivorous animals, which persecute the Moth, but avoid the Wasp. May not the Heliconidae dress serve the same purpose to the *Leptalis*? Is it not probable, seeing the excessive abundance of the one species and the fewness of individuals of the other, that the Heliconidae is free from the persecution to which the *Leptalis* is subjected?

* The Heliconidae include *Heliconius* butterflies proper and the Ithomiae. They share similar characteristics.

The gaudy colours, slow flight and carefree abandon of the Heliconidae suggested that they had no fear of predators. Many other butterflies of the Amazon succumbed to insectivorous predatory birds, but not the Heliconidae. Also, they were very strong-smelling: one species was reckoned to smell like fried rice, another like witch hazel. If the body of an *Ithomia* or *Heliconius* butterfly was broken, a strong-smelling juice exuded that irritated the skin. Bates had also noticed that his Heliconidae specimens were not attacked by vermin the way other butterflies were and that, when he encountered piles of butterfly wings on the forest floor, victims of predatory birds, the Heliconidae were not among them.

Bates reasoned that the Heliconidae were protected against predators by their unpleasantness. They acquire this toxic nature by feeding exclusively on species of the passion flower plant (species of *Passiflora*),* which itself produces cyanide-generating toxins as a protection. So the Heliconidae are exploiting a pre-existing defence mechanism and turning it to their own use, and the perfectly palatable *Leptalis* are getting a free ride on the back of the toxicity of the Heliconidae. This was the process which from now on would be known as Batesian mimicry: a defenceless creature obtaining relief from predators by assuming the pattern and colouring of one defended in some way, usually by inedibility – that is, by sailing under false colours. Bates admitted that this did not explain the fact that some Heliconidae appear to mimic other Heliconidae – all of them presumably defended against predators through their unpalatability. Here, Bates had missed a trick, because this kind of mimicry does have an explanation.

Bates's observations opened a subject which continues to develop to this day. He had discovered that natural selection was leading to a pattern's being copied from one species to another. What sort of biological process could have produced this? The quotation from Bates that heads this

---

* The passion flower is so named because the Spanish explorers who first found it in South America likened the structure of the flower to elements of Christ's crucifixion. The flowers have a very striking geometry, with five stamens and three large, spreading styles with flattened heads – these were taken to symbolise the wounds and the nails respectively. Although they are food for the caterpillars of *Heliconius* butterflies, it is bees that fertilise the flowers. Indeed, they seem to regard the butterflies as a nuisance, because they have evolved some amazing mimicry of their own. To deter *Heliconius* butterflies from laying their eggs on the plants, some passion-flower leaves have small yellow growths which mimic *Heliconius* eggs. A butterfly seeking to lay eggs needs a fresh leaf: seeing eggs already there, she will move to another leaf.

chapter reveals a paradox. Bates said: 'I obtain a glimpse of an intelligent motive pervading nature.' But biologists do not believe that one species consciously 'copies' another, or that nature has an intelligent motive. Such pattern making is intrinsically thrilling, but nowhere is there anything that could be said to have a motive. What survives does so by default, as every creature strives to stay alive and to reproduce. Beyond that individual striving there is no positive motivation. Bates understood this but tried to put a positive complexion on the workings of natural selection:

> It is clear, therefore, that some other active principle must be here at work to draw out, as it were, steadily in certain directions the suitable variations which arise, generation after generation, until forms have resulted which . . . are considerably different from their parent as well as their sister forms. This principle can be none other than natural selection, the selecting agents being insectivorous animals, which gradually destroy those sports or varieties that are not sufficiently like Ithomiae to deceive them.

Bates wrote up his Amazonian researches with Darwin's book to hand. The result was a classic paper, read at the Linnean Society on 21 November 1861 and published in the next year. Darwin wrote to Bates praising his paper for solving 'one of the most perplexing problems which could be given to solve', and produced an anonymous review for the *Natural History Review* in which he painted nature as a conjurer:

> We can understand resemblances, such as these, by the adaptation of different animals to similar habits of life. But it is scarcely possible to extend this view to the variously coloured stripes and spots on butter-flies; more especially as these are known often to differ greatly in the two sexes. Why then, we are naturally eager to know, has one butterfly or moth so often assumed the dress of another quite distinct form; why to the perplexity of naturalists has nature condescended to the tricks of the stage?

This was a rhetorical question, because Bates had already answered it. Happily, for Darwin the answer was – as Bates had said – that 'the butterfly or moth . . . assumed the dress of another' to gain the protection that accrued from looking like an unpalatable creature; and the agent of this process was natural selection. In the fourth edition of the *Origin*, Darwin added a section on Batesian mimicry, concluding: 'here we have an excellent illustration of natural selection.'

In the paper, Bates relates butterfly mimicry to many other adaptations, in which one species copies some part of the environment, or some other creature. Since Bates's time there have been some who maintained a hard and fast distinction between the strict imitation of one species by another and resemblances of other kinds. But, for Bates, these were all Darwinian adaptations in which one form had under selection come to assume a new superficial appearance because there was an advantage for survival in doing so – as in the example of moths, ants and other insects that imitate wasps:

> I believe therefore that the specific mimetic analogies exhibited in connexion with the Heliconidae are adaptations – phenomena of precisely the same nature as those in which insects and other beings are assimilated in superficial appearance to the vegetable or inorganic substance on which, or amongst which, they live.

Bates went on to list some of these adaptations, including 'the most extraordinary instance of imitation I ever met with': a large caterpillar (probably the elephant hawkmoth) which, when frightened, could puff up its face to resemble a snake and thereby hope to frighten off a predator.

Besides mimicry, what excited Darwin about Bates and the butterflies of the Amazon was the amount of variation Bates had observed. Both men were convinced that evolution was currently proceeding in a dynamic way among the butterflies of the Amazon. In March 1861 Bates had sent Darwin a paper on variation and the formation of species. Darwin commented: 'They seem far richer in facts on variation, especially on the distribution of varieties and subspecies, than anything I have read.' Bates replied: 'I am quite convinced that insects offer better and clearer illustration of the problems you occupy yourself with than any other class of animals or plants. It is so easy with them to obtain great series of examples and have them before you in a small compass.'

In the book he later wrote about his Amazon travels, *Naturalist on the River Amazons*, Bates made large claims for butterflies as model species for study:

It may be said, therefore, that on these expanded membranes nature writes, as on a tablet, the story of the modifications of species, so truly do all changes of the organization register themselves thereon. Moreover, the same colour-patterns of the wings generally show, with great regularity, the degrees of blood-relationship of the species. As the laws of nature must be the same for all beings, the conclusions furnished by this group of insects must be applicable to the whole organic world; therefore, the study of butterflies – creatures selected as the types of airiness and frivolity – instead of being despised, will some day be valued as one of the most important branches of Biological science.

Among the examples of variation that so excited Darwin, Bates found many intermediate wing patterns between two *Heliconius* species – *H. melpomene* and *H. thelxiope*. He rejected the idea that these intermediate forms were hybrids of the two species, because in places the two co-existed and did not seem to interbreed. Bates concluded, after reading *On the Origin of Species*, that he was observing the kind of variation that leads to a new species being formed. In some locations the variations became more successful than the original, eventually hardening into a new, set species. Bates expanded on the process:

It sometimes happens, as in the present instance, that we find in one locality a species under a certain form which is constant to all the individuals concerned; in another exhibiting numerous varieties; and in a third presenting itself as a constant form, quite distinct from the one we set out with. If we meet with any two of these modifications living side by side, and maintaining their distinctive characters under such circumstances, the proof of the natural origination of a species is complete: it could not be much more so were we able to watch the process step by step. It might be objected that the difference between our two species is but slight, and that by classing them as varieties nothing further would be proved by them. But the differences between them are such as obtain between allied species generally. Large genera are composed, in great part, of such species; and it is interesting to show how the great and beautiful diversity within a large genus is brought about by the working of laws within our comprehension.

Darwin reacted to Bates's findings: 'Whilst reading and reflecting on the various facts given in this Memoir, we feel to be as near witnesses as we can ever hope to be of the creation of a new species on this earth.'

In fact, the enormous variability of *Heliconius* butterflies, particularly *H. melpomene*, is still exercising biologists. Hybridisation is thought to be more important than Bates realised, and his *H. thelxiope* is currently regarded not as a full species but as a variety of *melpomene*. This is no discredit to Bates: the complexity of *Heliconius* is one of biology's great mysteries.

The early supporters of Darwin formed a tight little club in the face of the hostility of many members of the zoological establishment. Bates quickly became one of this inner circle, along with Wallace, Joseph Hooker and Sir Charles Lyell. Darwin wrote to Hooker: 'It is really curiously satisfactory to me to see so able a man as Bates (& yourself) believing more fully in natural selection, than I think I even do myself.'

Darwin became Bates's patron, encouraging him to write up his travels as a book and helping him to publish with his own publisher, Murray. Darwin reviewed *Naturalist on the River Amazons* when it appeared in 1863 and was unstinting in his praise. The book sold well and is a classic of travel literature. Bates could write, and I have a feeling that he would have written even better if it were not for the prevailing rules of Victorian decorum. In letters Bates was able to express himself with a rollicking gusto, far preferable to the measured sobriety generally favoured at the time. When the book appeared, he wrote to Darwin: 'It is most curious; all find pleasure in the book, Darwinians, Calvanistic [sic] church ministers, Dissenting parsons, hard-headed men of business; women, old men and boys; philosophic naturalists & species grubbers.'

CHAPTER 2

SWALLOWTAILS AND AMAZON

Suppose that a blue-eyed, flaxen-haired Saxon man had two wives, one black-haired red-skinned Indian squaw, the other a woolly-headed, sooty skinned negress – and that instead of the children being mulattoes or brown or dusky tints . . . all the boys should be pure Saxon boys like their father, while the girls should altogether resemble their mothers . . . Yet the phenomenon . . . in the insect world is still more extraordinary; for each mother is capable not only of producing the male offspring like the father, and female like herself, but also of producing other females exactly like her fellow-wife, and altogether differing from herself.

Alfred Russel Wallace, 1865

Alfred Russel Wallace's Indonesian expedition (1854–62) ought to be as famous as Darwin's *Beagle* voyage. Not only did Wallace arrive at the same concept of natural selection as Darwin: two further major discoveries made the trip much more than mere collecting, although he did a lot of that too. Indonesia proved a happy hunting ground because two different sets of animals – Asiatic and Australasian – are found very close together but not mingling. The dividing line is caused by deep water, which formed a barrier to animal movement, including that of small birds. The line passes between the islands of Bali and Lombok, which are 35 miles apart. This was important evidence for evolution: yet another example of isolated populations evolving different species. The line is still called the Wallace Line.

Wallace's second big find was one of the most remarkable cases of mimicry, encountered on Sumatra in his first year: one which was to launch countless papers and is still going strong today. *Papilio* butterflies are the swallowtails, a family of over 500 butterflies found on every continent. Not every member of the family has the characteristic forked tail, and even within the same species some do and some don't – a paradox which lies at the heart of the mystery of mimicry. In Asia, Wallace saw male and female swallowtails so different in appearance that they had previously been supposed to be separate species. The males were large butterflies with a black background colour, deep blue coloration in parts and no tails. The females couldn't have been more different, being whitish with buff, yellowish and red markings, and some had tails. These females mimicked several other species, including a different kind of swallowtail. Some had tails, some did not. Wallace interpreted this as an example of Batesian mimicry. Most curious of all was the finding that the female of another species of swallowtail could produce two different females from a single brood: one which resembled the male and one which mimicked yet another species of swallowtail.

Swallowtails were also the fascination of another British naturalist on a different continent. Roland Trimen (1840–1916) left England for South Africa at the age of eighteen and, while working as a civil servant, became expert in the butterflies of the region, eventually publishing voluminous books on the subject. Once he had read Darwin and Bates, he came to believe both in natural selection and in the phenomenon of mimicry.

A sometimes hilarious episode, which has become known as 'The Case of the *Merope* Harem', began when, in the early 1860s, Trimen was puzzled by the swallowtail butterfly *Papilio merope*: no females could be found. After reading Bates, Trimen began to suspect that he was witnessing Batesian mimicry. Four 'species' came under suspicion of supplying 'wives' for *P. merope*, although they were very different in appearance. When Trimen came to England in 1867, to research the butterfly collections in the British Museum, he could find not a single female of *P. merope* and no male for any of the suspected 'wives'. Then he realised that they must be the males and females of the same species: *P. merope*. Trimen presented his findings to the Linnean Society on

5 March 1868. The anti-Darwinian W. C. Hewitson, a noted butterfly collector with the largest collection in England, objected in a wonderfully stuffy Victorian manner: 'it would require a stretch of the imagination, of which I am incapable, to believe that the *P. merope* of the mainland, having no specific difference, indulges in a whole harem of females . . .'. The idea shocked his 'notions of propriety'. As an objection, he remarked that on Madagascar there were *P. meropes*, both male and female, and they were identical. This was true – things were different on Madagascar – and this anomaly would not be explained for one hundred years.

Wallace had reported multiple forms of Asian swallowtails from one brood. What was needed was proof by breeding that the multiple forms could all come from one parent. Frustrations ensued but eventually, in 1874, J. P. M. Weale bred male *meropes* and three of the females from a single brood. Later broods produced the final female form. So the composition of the *Papilio merope* family was established at last.

But why would a single species of *Papilio* need to have multiple mimetic models? The answer is ingenious and inherent in the nature of Batesian mimicry. To survive, Batesian mimics needed to be scarcer than their models. If they were numerous, predators would encounter many palatable as well as unpalatable butterflies with the same appearance and they would start attacking both the models and their mimics. But by imitating several different species these swallowtails spread the load, as it were, and could increase in number without endangering the protection that mimicry afforded. So, when they emerged from their chrysalises, the little tribe of polymimetic swallowtail butterflies hit the ground running, being well prepared for life in their different costumes. And all of them came from a single batch of eggs.

Weale's paper on the breeding behaviour of *Papilio merope* is amusing and gives a good insight into the world of (some) British naturalists of the nineteenth century. He writes of the courtship of the butterflies in a style which seems to come from pulp romantic fiction:

Her mate is not generally so early on the wing, but shortly afterwards he may be seen hurriedly darting over the bush, down on some flower, then up again and away. At this hour he seems to pay little attention to his lady-love.

As the day grows warmer, the females generally, but not always, glide away into some shady spot, often settling for long periods, or occasionally gliding about in their cool and sequestered bowers. The males at this time chase each other in a rapid and violent manner, constantly passing and repassing the hidden nook, where their lady-love has coyly retired.

Commenting on Weale's paper, Trimen twitted Hewitson:

It is with reluctance that one contemplates the stretching of Mr. Hewitson's imagination to an extent 'of which he is incapable', or the inevitable shock which his 'notions of propriety' will receive, but the evidence now adduced by Mr. Weale is such that the profoundest sceptic cannot explain it away, and must allow that the dream had proved to be a true vision.

It is obvious that some nineteenth-century British naturalists such as Hewitson could not view nature without projecting onto it coyness and gentility. Darwin's tough-minded stance shines out as a beacon throughout all this, and his achievement appears all the more striking when one knows the climate in which he operated. Although British naturalists were very prominent in the Victorian era, the *Zeitgeist* was in some ways inhospitable to scientific enquiry. Bates shared Darwin's rigour, but Wallace eventually lapsed to some degree, sliding into spiritualism and denying that the human race was entirely a product of evolution. For many Victorians, leaving their Christian comfort zone was just too much. An honest attempt to wrestle with these difficulties can be seen in Tennyson's poem 'In Memoriam'.

Completed in 1849, ten years before *On the Origin of Species*, 'In Memoriam' reads today almost as a commentary on the shock experienced by Victorian morality when confronted by the harsh Darwinian lesson of the cruelty of nature:

Are God and Nature then at strife,
    That Nature lends such evil dreams?

So careful of the type she seems,
So careless of the single life;

. . .

'So careful of the type?' but no.
   From scarpèd cliff and quarried stone
   She cries, 'A thousand types are gone;
I care for nothing, all shall go.'

To return to Weale: he had succeeded in raising the four different females and the male from pupae, but he was never actually able to catch the males and the dissimilar females in the act of mating. The final proof had to wait another seven years, when a male *Papilio merope* and one of the females were caught *in flagrante*. The butterfly is no longer called *Papilio merope* but *Papilio dardanus*. It is very celebrated, has around thirty different forms, and will flit in and out of our story.

But why is it only female swallowtails that display this extravagant mimicry? Their males remain constant in form. This question has exercised many. There are plausible answers: female butterflies are especially vulnerable to predators when they are laying eggs, so they need the extra protection; or, the females select for constant mate patterns in mate choice. Several solutions have been proposed, but there is still no definitive answer.

Bates, Darwin and Wallace were eminent Victorians, members of a noble breed. Their circumstances had been very different: Darwin had private means, the benefit of a Cambridge education, and contacts among the scientific illuminati; Bates and Wallace had very humble beginnings, were truly self-educated, and had struggled to support themselves on their expeditions. But, as naturalists, all three were of a piece: they had followed their passion, from childhood beetle collecting to their adult quest for the origin of species, without recourse to received wisdom and authority. They were independent thinkers and they treated each other with great courtesy. When they were in town, Bates and Wallace attended meetings of the Zoological, Entomological and Linnean Societies. Wallace never resented Darwin's claim to be the principal originator of the theory of natural selection. On Christmas Eve 1860, Wallace wrote to Bates one of the most magnanimous letters anyone has ever written:

I know not how, or to whom, to express fully my admiration of Darwin's book. To him it would seem flattery, to others self-praise; but I honestly believe that with however much patience I had worked and experimented on the subject, I could never have approached the completeness of his book, its vast accumulation of evidence, its over-whelming argument, and its admirable tone and spirit. I really feel grateful that it has not been left to me to give the theory to the world.

With the *Origin* behind him, Darwin began to work on a book about human evolution, eventually published in 1871 as *The Descent of Man*. During the course of writing this book, he pumped his friends Bates and Wallace for information. An important aspect of the book was its concern with sexual selection, an agent of evolution distinct from natural selection and complementary to it. Darwin reasoned that, if there is competition among males to mate with females, this must have an influence on the composition of future generations. He attributed the evolution of bright colouring in many males (when the female was often quite dowdy) to this process of sexual selection. In butterflies, coloured wing patterns can serve several functions: besides mimicry and camouflage, they are concerned with sexual selection; and there are startle patterns and eyespots to lure predators away from the vulnerable head. One aspect of bright colouring especially worried Darwin. Sexual selection could obviously only occur in adult creatures that mated. But many caterpillars – the immature, non-mating form of butterflies – also exhibited lurid colouring. On 23 February 1867 Darwin wrote to Wallace, asking: 'Why are caterpillars sometimes so beautifully and artistically coloured?' He had already asked Bates, who could find no answer and suggested: 'You had better ask Wallace'.

At the time, Wallace was writing a paper on 'Mimicry and protective colouring' for the *Westminster Review* which was a very good synopsis of the state of the art, and he immediately had the idea that it might not be enough for a caterpillar to be poisonous or distasteful to save itself from a predator. By the time the predator has discovered just how unpleasant a meal the thing would be, enough damage might have been done to kill it. So, in this case, distastefulness would be no protection at all. The unpleasantness needed to be advertised in as strong a manner as possible. Hence the gaudy colouring.

As it happened, Wallace had just discovered an example of a *white* moth which was avoided by birds. As he noted in his autobiography: 'Now, as a *white* moth is as conspicuous in the *dusk* as a *coloured* caterpillar in the *daylight*, this case seemed to me so much on a par with the other that I felt almost sure my explanation would turn out correct.'

Darwin was delighted, replying:

> My dear Wallace, – Bates was quite right, you are the man to apply to in a difficulty. I never heard anything more ingenious than your suggestion, and I hope you may be able to prove it true.

Wallace strangely missed a trick here. Although the white moth helped him to formulate his theory about the warning colouring of toxic caterpillars, he did not at the time generalise it to wonder why so many toxic butterflies were brightly coloured. Some chapters in the history of mimicry seem to have been written in the wrong order. But there the story of mimicry rested until 1879. The three naturalists remained mutually supportive adherents of the theory of natural selection, although – inevitably, as the years passed, given that evolutionary biology was in its infancy – differences began to emerge. Wallace at one point in his autobiography summed up his differences with Darwin and hindsight has sometimes proved him correct. Darwin believed that the gaudy colours of many males of many different kinds – from butterflies to peacocks – were the result of sexual selection. Wallace eventually came to believe that most bright colours were examples of warning coloration. In many cases, modern work has sided with Wallace.

Darwin made much of the concept of disuse in *On the Origin of Species*. Moles, for instance, through countless generations of living under ground, have become virtually blind. The facts are incontrovertible – moles are blind, but early moles were mammals with eyes like any other (in the twenty-first century we can be sure of this because the eye genes are still there, battered and holed by mutation but still recognisably eye genes). But Darwin somewhat abandoned his rigour in considering how this may have come about. He was at times tempted by the theory of the French naturalist Jean-Baptiste Lamarck (1744–1829), the first systematic thinker on evolution, who in 1800 proposed that characteristics inherited during life could be passed on to the offspring. This idea has been a persistent heresy because to some minds it is highly appealing. In *The Variation of Animals and Plants under Domestication* (1868), Darwin clearly claims that acquired traits can be inherited: 'A horse is trained to certain paces, and the colt inherits similar consensual movements.' Wallace, on the other hand, rejected Lamarckism firmly and entirely.

But the most important difference between Darwin and Wallace was the latter's refusal to believe that evolution could account for all of the attributes of human beings: 'I, though agreeing with him [Darwin] with regard to man's physical form, believed that some agency other than natural selection, and analogous to that which first produced organic life, had brought into being his moral and intellectual qualities.' In other words, God had started the process of life, which then continued under the process of evolution by natural selection, until the emergence of Man, when God intervened again, to add a sprinkling of moral and intellectual qualities to the otherwise ape-like evolved form. Bates referred to Wallace's 'backsliding' on this question.

Bates did not 'backslide' on the question of natural selection, but his career following his great paper on mimicry and his travel book was strangely muted. Without Darwin's private means, and needing permanent employment, he had set his sights on a job with the British Museum's natural history collection. He was the best qualified person in the country for this job, but Victorian patronage saw the matter differently. The museum authorities wished to appoint a favourite, a young poet with no botanical or zoological qualifications: Arthur O'Shaughnessy. This was a scandal and some of Bates's friends protested, but to no avail.

With this disappointment behind him, Bates secured the post of assistant secretary at the Royal Geographical Society. He continued to write papers on the classification of his beloved beetles and built an excellent reputation for assisting explorers embarking on just the sort of trip he had once made. There were no more travels or ground-breaking papers for him, but the butterfly mimicry paper remains a great achievement and is still often cited. As a paper announcing a major new theme in biology, it could not be bettered.

There is one more twist to the tale of butterfly mimicry in the nineteenth century. As we saw, Bates had noticed that, besides palatable species imitating unpalatable ones to gain protection, many species of *Heliconius* – all unpalatable – mimicked *each other*. Having discovered the elegant process now known by his own name – Batesian mimicry – he was understandably reluctant to admit that this form of imitation was truly mimicry at all. Given that it occurred within related species, he attributed the phenomenon to parallel variation or to 'similar adaptation of all to the same local, probably inorganic conditions'.

The next step was taken by a German (strictly speaking Prussian, since Germany did not yet exist as a nation–state) zoologist, Fritz Müller (1821–97). Müller was an unusual character, very unlike the oh-so-Victorian figures of Darwin, Bates and Wallace. He was a political radical and an atheist.

Müller's background showed no signs of the independence to come. His father and grandfather were churchmen and his maternal grand-father was a distinguished pharmacist. Müller was well educated, loved maths and, like Bates, Darwin and Wallace, had been a passionate naturalist from an early age.

But he was young at a time of great ferment in Europe. At university, he became a religious and political radical just when religious orthodoxy was demanded of people in public life. Following the failed revolution of 1848, German society became ever more repressive and Müller realised that he would not be able to live in such a world. His response was to drop out, abandoning hopes of a university post, abandoning also his childhood sweetheart (whose family would never have accepted him). He began to live with an uneducated labourer's daughter. In 1852, Müller emigrated to Brazil with his partner and young children (he even-tually had six daughters). He did then marry, but only because he had been warned that his partner's position would have been intolerable in Brazil without it.

Blumenau was a small German-speaking colony in south-eastern Brazil. There Müller lived like a good proto-hippie, clearing the land, building a homestead, farming, and fighting off attacks from the indige-nous population, Wild-West style. He was proud of having carved out a homestead in the jungle and of providing for his family by his own labour. A photograph of him in later life shows a lanky figure with long straggly hair and beard.

Müller was, above all, a scientist as omnivorous and questing as Darwin himself. He was one of Darwin's early supporters; but, whereas Darwin admitted that he had no aptitude for mathematics at all, Müller, after four years of full-time homesteading, was offered a post teaching mathematics in the provincial capital. This gave him the opportunity to reclaim his scientific work. In 1861 he read Darwin's *Origin of Species* and, after wrestling with its ideas, became one of the most staunch Darwinists. He resolved to test these ideas in his researches. This resulted in a book of advocacy, *Für Darwin*, published in 1864. When Darwin received this he was delighted. So began a long correspondence, and Müller became one of Darwin's closest colleagues. Darwin arranged for the translation and publication of *Für Darwin* in England. Through Darwin's good offices, many of Müller's papers were published in England in the leading journal *Nature*.

Müller might have been isolated living in a German colony in remote Brazil: the heartland of entomology was undoubtedly England. Although German zoology was very distinguished, if his ideas were to be accepted, Müller would have to be heard by the very English club of gentleman naturalists. But Müller had some advantages. Part of his brilliance was his gift for languages: his English was almost perfect.

In 1876 Müller was appointed official naturalist to the National Museum in Rio. This enabled him to travel throughout Brazil and to pursue his researches more fully. In 1870 Darwin had sent him the German translation of Wallace's *The Malay Archipelago*, and the two exchanged views on mimicry. For some years Müller entertained a novel idea, designed to account for the many mimetic butterfly species which shared both unpalatability and similarity of appearance (unlike in Batesian mimicry, where one palatable and one unpalatable species shared the same appearance). It was, according to Müller, 'not improbable that a butterfly endowed with a highly developed sense of colour found the pretty males or females of another species more to its taste than those of its own species'. And so, through this preference, the likeness might be achieved.

Müller clung to this idea for several years, but the true explanation had been suggested to him by Darwin himself, in August 1870. Having recently learnt from Wallace that the bright colours of unpalatable caterpillars are warning colours, Darwin proposed that 'The same view perhaps applies in part to gaudy butterflies'. It took Müller over seven years to become convinced of this but, as he said afterwards, 'It is remarkable how one racks one's brains over a problem the solution to which is so simple that it is scarcely conceivable how one could have found it so difficult a moment before'.

In February 1878 Müller encountered two species similar in appearance, which he recognised as the unrelated *Ituna ilione* and *Thyridia megisto*. They had very different vein structures on the wings, but visually they matched. The butterflies were clearwings, that is, they were partially transparent, with whitish patches on the forewings and a sepia see-through tint on the hindwings. Both were also unpalatable. Müller realised quickly that his sexual selection idea was wrong.

Then Müller went way beyond Darwin's hint. His argument was ingenious and somewhat mathematical. He reasoned that, however well established the connection between warning coloration and distasteful prey may have been, young birds would need to learn this lesson the hard way. In the process, some individual butterflies would be killed. The price of the enhanced survival of the species was this toll on individual butterflies.

On the other hand, if several unpalatable species shared the same pattern, the chance of any one species losing many individuals would be drastically reduced. And, proportionally, the gain would be much higher for the less numerous species. In one of the first applications of mathematics to biology, Müller showed that, if a species was, say, five times less numerous than the other, the loss from predation would be reduced by 25 times by comparison with the more numerous species: the gain is the ratio of the two populations squared (in this case 5^2). The pressure for an enhanced survival rate of the less numerous species would be a natural selective force driving that species into better mimicry with the more numerous model. This process has then continued until a very good match has been achieved between the two species.

Müller's idea explained the existence of mimicry rings: groups of several species which share the same patterning, even though some members of these groups are not even butterflies, but day-flying moths. It is an insectivorous protection racket. When a bird sees the standardised warning pattern on the wings, it doesn't need to know what kind of creature it is – it is to be avoided, and that is all it needs to know. The incentive for an insect to join such a club was irresistible.

But why, then, if several species have clubbed together to pool their protection, has not the process gone to completion; which would mean that all of the unpalatable butterflies in one place would share the same patterns? This is very far from the case: where there is one mimicry ring there are usually five, each with a different pattern, operating in slightly different habitats – at different heights of the forest canopy, for example – but still to some extent overlapping. This is a continuing puzzle. The most likely explanation is that each mimicry ring inhabits a distinctive microhabitat.

In 1879, Müller's new form of mimicry was communicated to the Entomological Society in London. Sadly, Bates was not able to rise to the occasion. He 'could not see that Dr Müller's explanations and calculations cleared up all the difficulty'. Wallace, too, was sceptical of the new form of mimicry, believing that the resemblance between two unpalatable species might be due to 'peculiar elements or chemical compounds in the soil, the water, or the atmosphere'. But in subsequent correspondence Wallace came to realise the force of Müller's arguments and in 1882 he published in *Nature* a glowing acknowledgement of the new form of mimicry. Müllerian mimicry – in which unpalatable species club together to wear the same dress, putting up a united front recognised by all potential predators – was born.

In Darwin's later years, Müller was one of his most regular and trusted correspondents. Two years before Darwin's death, he wrote of Müller: 'I have long looked on him as the best observer in the world.'

The discovery of mimicry was associated with a great revolution in biology: the theory of natural selection. Understanding how the process worked required several more revolutions, equally profound. The most pressing question concerned inheritance: how do creatures produce offspring which are very like themselves but different in subtly significant ways? The solution to this problem was actually already known soon after Darwin published the *Origin*, but the work which propounded it remained unnoticed until 1900.

In the meantime, the great explorations continued. A panoply of mimicry opened up, involving many bizarre resemblances – not just in butterflies, but in spiders, ants, mantises, snakes, sea dragons. Much of the animal word, it seemed, was driven by forgery.

CHAPTER 3

DELIGHT IN DECEPTION

Every naturalist traveller appears to have some instance to relate of how he was taken in by a protectively-coloured insect. The stories are told with a curiously exaggerated delight at the deception, and often with a framework of details tending to throw the deception into still greater prominence . . .

F. E. Beddard, *Animal Coloration* (1892)

The half century between Bates's paper on mimicry and the outbreak of the First World War saw a golden age of discovery, especially by British explorers. In the reports these travellers brought back from Asia, South America and Africa there were many remarkable examples of mimicry and camouflage. The observations piled up, but there were no new insights into the how and why of nature's penchant for copying to add to the concepts of Bates, Wallace, Darwin and Müller.

Out of the myriad examples of mimicry, a few take pride of place for their sheer élan. One such was discovered by the naturalist and explorer, my namesake Henry O. Forbes (1851–1932), who travelled in the Malay Archipelago, very consciously in Wallace's footsteps, from 1878 to 1883. Forbes was a Highland Scot who had lost an eye while studying medicine and had turned to exploration. In March 1878, on the island of Java, he was pursuing a butterfly through prickly *Pandanus* scrub, when it came to rest on what seemed to be a bird's droppings on a leaf. Forbes was able to pluck the butterfly from the leaf, and at first thought that part of it had remained stuck to the excreta.

There *was* an insect on the leaf – but it was not part of the butterfly. Bird droppings are ubiquitous objects in nature: chalky white streaked with black, with a fluid portion, partly dried up but still glistening in places, running down the leaf. Insects, including butterflies, are attracted to excrements for the salts they contain. Several creatures mimic bird droppings, especially caterpillars. In that case, they would be using the disguise to hide from predators but spiders might wish to prey on insects attracted to bird droppings. They have an advantage in achieving mimicry: the web. In some black-and-white spiders, the web has been transformed into a film which mimics the runny part of the excreta. This is what Forbes had stumbled upon: a bird-dropping spider.

Later in his travels, in Sumatra in June 1881, Forbes encountered the same species again, once more by accident. He became aware of a leaf marked with a bird's droppings. This triggered the thought: 'How strange it is that I have never found another specimen of the curious spider I got two years ago in Java, which simulated a mark just like this!' He plucked the leaf half-heartedly, still thinking about the deceptive spider he had seen previously, when the thought spider suddenly became a real spider again.

Forbes was struck by the fact that, although bird droppings have some fairly constant features, they are not as precise an object as a living creature. Although Forbes was a Darwinian, he felt his faith in natural selection challenged by the ability of a living creature to mimic accurately the appearance of something so shifting and elusive.

The patterns which have evolved on the bodies of insects are usually constant and, if one insect has copied another in mimicry, the model has stayed constant as the mimic has evolved. But bird droppings are variable in form – they are recognisable but not constant in detail:

> There is no doubt that the spider must have acquired this mimicking habit by natural selection; yet it is difficult to explain how these minutiae, which are not constant or essential in the model, have come to be so accurately copied; one cannot believe that it would have been one whit worse off had the copy been less minutely imitated.

But, interestingly, when Forbes sent his spider (now called *Ornithoscatoides decipiens* (a wonderful name: *ornitho* [bird]-*scatoides* [dung-like]) back to England, the entomologist Reverend O. Pickard-Cambridge (1828–1917) was brisk with Forbes's doubts about natural selection:

> It seems to me, on the contrary, that the whole is easily explained by the operation of natural selection, without supposing consciousness in the spider in any part of the process. The web spun on the surface of the leaf is evidently, so far as the spider has any design or consciousness in the matter, spun simply to secure itself in the proper position to await and seize its prey. The silk, which by its fineness, whiteness, and close adhesion to the leaf causes it to resemble the more fluid parts of the excreta, would gradually attain those qualities by natural selection . . .

It is good to see a Reverend taking the hard-nosed scientific line. In fairness to Forbes, Reverend Pickard-Cambridge included a footnote acknowledging that Forbes, having read his critique, had 'expressly disclaimed the idea of crediting the spider with any conscious design', but Forbes went on: 'the similitude is so exact, that the spider *might have* had consciousness, and it could not have been more exact if the spider did have it'. Forbes was clearly struggling to maintain his grip on the problem, but he concludes, rather triumphantly: 'Is not its exactness probably the result of the *un*consciousness of the spider? Conscious design would possibly have resulted in failure and abandoning the plan, or at best in a more clumsy imitation.'

In this example we can see many of the dilemmas of mimicry. It is so easy to slide into disbelief, but Forbes in the end hits on one of the great truths of nature. When we catch a ball we are conscious of what we do, but we are not *consciously making the calculations* that the brain and nerves must make unconsciously. It is human vanity that refuses to acknowledge the power of natural processes, unmediated by human consciousness. In the case of the constructions of nature such as spiders' webs, imitation bird droppings, birds' nests or any structures built by an animal, the behaviour is the result of a genetic programme ('instinct', if you like) and not of conscious choice.

Besides being vain, we tend to be squeamish and prudish. It may seem strange to find beauty in something which mimics excreta, but the elegance of this contrivance is surely beautiful in the old alchemical sense of *aurum ex stercore* – gold from filth, the rose emerging from rotted manure, the pure lotus flower casting off mud in the murky waters. Primo Levi (1919–87), the renowned chronicler of the Holocaust and prophet of a unified culture embracing both art and science, caught this mood in *The Periodic Table* when he was attempting, in post-Second World War poverty-stricken Italy, to make chemicals for lipstick from chicken dung: 'The trade of chemist ... teaches you to overcome, indeed to ignore, certain revulsions that are neither necessary nor congenial: matter is matter, neither noble nor vile, infinitely transformable, and its proximate origin is of no importance whatsoever.' Similarly, a leaf form or a stone does not have to be made from plant matter or mineral rock to have the appearance of one of these objects. Some of the most remarkable mimetic creatures – for example the leaf butterflies, of which the Indian leaf butterfly (*Kallima inachus*) is the finest, or the living stones of South Africa – show creatures subverting the conventional shorthand animal/vegetable/mineral classification. The leaf butterfly is animal tissue pretending to be plant; the living stones plant pretending to be mineral.

Living stones were first reported by an early nineteenth-century traveller to Africa, William Burchell (1781–1863), and the interest in mimicry in the 1890s brought them back into focus. Burchell's encounter with the stones in southern Africa between 1810 and 1815 follows the well-worn pattern of exaggerated delight in deception noted in the epigraph to this chapter:

> On picking up from the stony ground, what was supposed a curiously shaped pebble, it proved to be a plant, an additional new species to the numerous tribe of Mesembryanthemum; but in colour and appearance bore the close resemblance to the stones, between which it was growing ...

In 1900, a German naturalist, Karl Dinter, travelling in the region, also recognised the living stones as 'a very characteristic form of self-protection, or true mimicry'. When not in flower, the 'stones' are leafless and their texture is pure stone, but in bloom they sprout the familiar daisy-like *Mesembryanthemum* flower.

Leaves and stones are mimicked by several different kinds of creatures: there are the *Pêche de Folha* leaf fishes of Brazil; the stone fishes, which use their brilliant camouflage to attack with highly poisonous barbs; and there are many seaweed mimics, the leafy sea dragon being the finest: an apparently careless stringing together of seaweed fronds.

Wallace had drawn attention to the leaf butterfly, *Kallima*, a genus found throughout south and east Asia, as one of the most perfect examples of camouflage. The upper surface of the wings is richly coloured in indigo and brownish orange. But when the butterfly alights on a leaf, closing its wings, it looks astonishingly like a dead leaf. It assumes the form and position of a dead leaf, and it even seeks out plants with dead leaves in order to blend in with them.

Kallima has always been a *cause célèbre*. It was lauded by Darwin: '[It] disappears like magic when it settles in a bush; for it hides its head and antennae between its closed wings, which, in form, colour and veining, cannot be distinguished from a withered leaf together with the footstalk.' But there has been a persistent streak of criticism that *Kallima* is *too* perfect; it began with the American palaeontologist Richard Swann Lull (1867–1957), who wrote in 1917: 'a much less perfect imitation would be ample for all practical purposes and we cannot conceive of selection taking an adaptation past the point of efficiency'.* This argument would rage for decades to come.

Kallima only does an impression of a single leaf, but some insects collaborate to form a collective plant-like structure. The most startling example of this phenomenon was discovered by the British geologist and geographer J. W. Gregory (1864–1932) during his explorations in East Africa in the 1890s. Gregory was walking in woods near the Kibwezi River when he noticed what appeared to be a bright flower spike. What really caught his attention was some unusual white fluffy patches below the flower. These resembled lichen and lichen doesn't normally grow on

---

* Lull believed that besides natural selection there was a process he called genetic drive: on this theory, the Irish elk had become extinct because its antlers grew unstoppably big through genetic drive.

flowers, so Gregory investigated. He pushed his stick through the bushes to pull the inflorescence closer, when the 'flowers' all rose and flew away. Some of the 'flowers' seemed to be pink, others greenish. They turned out to be a plant-juice-sucking species of insect: *Ityraea nigrocincta*. The white fluffy 'lichen' consisted of larvae, and the flower-like forms were the adults.

We are always impressed by collective action in nature, whether it is the geometrical precision of the bees' honeycomb or the team spirit of meerkats standing on their hindlegs as lookouts for the colony. That natural selection should have corralled these sluggish insects into a passable imitation of a flower spike is just as remarkable. Their sluggishness is a clue. They do not fly as well as some other members of their genus, so *Ityraea*'s strategy seems to be to sit tight on a stem, hoping for a quiet life. If a predator such as a bird rumbles their disguise, they hop off the stem and hope that some at least will fall on places where they would be hard to catch.

So far so good; but, as with so many stories of the natural world, this one soon became complicated and uncertain. Other observers could not entirely corroborate Gregory's findings. He had proudly placed an illustration of his discovery as the frontispiece to his book *The Great Rift Valley*, reproducing the flower spike as he had seen it. The insect exists in two forms: greenish and orange-red, and Gregory believed that the orange-red insects assembled on the lower part of the stem with the greens at the top, cleverly mimicking a flower spike with the topmost buds yet to open. It is not wise to draw conclusions about anything in nature on the basis of one example, but this is what Gregory did. It seems that the insects aren't that clever – the two forms are normally simply mixed up on the spike. There is no slowly maturing flower spike.

The case of *Ityraea* reminds us that insects and flowering plants co-evolved and that their interactions go way beyond the basic 'contract' – flowers attracting insects by colourful advertisement, rewarding them with nectar, and achieving cross-pollination in return. Flowers are often arranged, as Darwin was delighted to discover, to be pollinated by a single species of insect. In such cases the flower has evolved a shape which is inaccessible to all but one insect. The most dramatic example of this phenomenon was also one of the best confirmations of Darwin's theory. In 1862, working on the pollination of orchids, he was sent an exotic species from Madagascar, with an incredibly long nectar spur. Darwin predicted that a moth with a proboscis 20–35 cm long must be the pollinator. No such moth was known until 1903, when a hawkmoth with a proboscis six times the length of its body was found on the island.

But, just as some species such as Darwin's orchids are 'specialists', attracting only one species of insect, others are 'generalists', attracting many kinds. It is then possible for an insect to eavesdrop on this system and use the attracting power of the flowers for its own purposes. Insects such as the flower mantises, which lure their prey by disguising them-selves against a floral background, are some of the most intriguing mimickers.

In the popular imagination, praying mantises are famous for the females' habit of devouring the males after mating; but something far more noteworthy was seen by the Scottish zoologist Nelson Annandale (1876–1924) in Malaya in 1899:

> I was attracted to a bush of the 'Straits Rhododendron' (*Melastoma polyanthum*) by a curious movement among the flowers of a large inflorescence at the height of about five feet above the ground. On a cursory examination I could only see that one of the flowers – so it appeared – was swaying slowly from side to side; and it was not for several seconds that I realised that the moving flower was not a flower at all, but a Mantis.

The mantis was the orchid (or Malayan) praying mantis (*Hymenopus bicornis*), and it can be found on many pinkish-red flowered plants. It matches the beautiful orchid colours of white, red and tinges of green in an uncanny way: insect beauty paying tribute to the floral kind. But its goal is less lovely – to lure and consume bees.

Since Annandale's observation, many different species of flower-mimicking mantises have been found. Some tailor their colour to the flowering season; insects that hatch in the early season before the flowers are fully out are greenish, the later ones pinkish.

With stories such as Forbes's spiders and Annadale's mantises, we should bear in mind that habitually exaggerated delight: the travellers' tales and exotic discoveries were satisfying for a while, but a deeper understanding than exclamations of surprise and wonder was called for. The mass of information on camouflage and mimicry cried out for a systematiser. After Bates, Wallace and Müller, the leading figure was the English entomologist Edward Bagnall Poulton (1856–1943). Unlike his predecessors, Poulton was not an explorer but a collator and analyst of mimicry and camouflage.

At Oxford University, at the age of twenty-three, Poulton read Wallace's essay on natural selection and this 'aroused a lifelong delight in the facts and theories of Protective Resemblance'. Poulton spent almost his entire life at Oxford, becoming Hope Professor of Zoology in 1893. He was Oxonian to the core and, although he lived well into the Second World War, his sensibility remained Victorian. He was a habitual writer of verse jingles, especially limericks. These mostly exhibited the coy silliness typical of the genre, and science rarely intruded. When it did, it sounded like this:

> Said the Scientist to the Protoplasm –
> 'twixt you & me is a mighty chasm:
> We represent extremes, my friend
> 'you the beginning, I the end';
> And the protoplasm made reply,
> As he winked his embryonic eye –
> 'Yes, and when I look at you, old man,
> 'I'm rather sorry I began.'

Admittedly this was in his retirement year, 1933; but his verse was never pitched higher. Although it was said of him by an old friend that 'he could forgive anything except disbelief in evolution', he retained a traditional Christian piety, writing out in his notebook little prayers to offer at the alumni club to which he belonged, 'The 80 Club'.

Poulton remained essentially a naturalist, with an almost artistic love of the patterns which mimetic creatures assumed. He had no interest in quantifying the phenomena he investigated. This distinction – between the observation of creatures in the wild, as recorded so lovingly by naturalists such as Bates, Wallace, Trimen, Forbes, Gregory and Annandale, and the statistical procedures and mathematical conceptualisations of the harder-nosed experimentalists and theorists – has continued to dog the subject of mimicry and camouflage. Anyone who observes a bird dropping suddenly revealing itself to be a spider or an elephant hawkmoth caterpillar suddenly adopting its snake's-head startle pose, or sees unrelated butterflies flying together and wearing almost precisely the same patterns, is certain that their eyes don't deceive them. There is no need for statistics. But eventually field observations, controlled experiments and correlations with genetics have to be reconciled. This is the task now, in the twenty-first century. In Poulton's time, controversy was bound to rage unchecked because biology was still too primitive a science to allow the deeper patterns, which lie beneath the surface appearance of these creatures, to be discerned.

In his 1890 book *The Colours of Animals*, Poulton reviewed most of the examples of protective, alluring and warning coloration then known. His analogies for Batesian and Müllerian mimicry are neat and pithy: 'A Batesian mimic may be compared to an unscrupulous tradesman who copies the advertisement of a successful firm, whereas Müllerian mimicry is like a combination between firms who adopt a common advertisement to share expenses.'

As the first to pigeonhole and tidy up the various types of mimicry and camouflage, Poulton introduced a clear system for differentiating the various ways of using deceptive visual appearances; and this system is still useful for keeping track of the various techniques. The following are Poulton's headings (each one accompanied by my own paraphrase):

- **protective resemblance – special and general**
  Special resemblance is exhibited when the whole creature has the appearance of a discrete object; leaf, twig, stick and stone mimics come into this category. *General* resemblance is exhibited when the texture of the creature blends with the background. Most examples of camouflage fall into this category, especially the bark mimicry of many moths.

- **aggressive resemblance**
  This can be special or general. A camouflaged predator which hides against a general background to surprise its prey is showing *general*

aggressive resemblance; a flower-mimicking predator such as the flower mantis displays *special* aggressive resemblance. There are two methods: a flower can simply hide the predator, or it can lure both the predator and its prey, as do the flowers on which crab spiders and bees are found.

- **adventitious protection**
This category refers to the use of props, for instance a creature that arranges material – grass, stones, twigs, or shells – on its own surface, to obscure its outline. Some crabs cover themselves in seaweed and stones, and caterpillars of camouflaged looper moths of the genus *Synchlora* snip off flower petals with their mandibles and stick them onto their back.

- **variable protective resemblance**
This is a kind of resemblance shown by higher animals such as flatfishes, some amphibians, octopuses and squids, that can change their colour to match their surroundings.

- **warning colours**
Creatures which are protected by toxins that make them unpalatable generally advertise the fact through the use of gaudy coloring, usually in red, yellow, black or white. Wasps, ladybirds, *Heliconius* butterflies and coral snakes are prime examples.

- **protective mimicry**
One creature copying another more dangerous one, as in Batesian mimicry; or creatures protected by chemical or other defences banding together under the same coloration, so that they all benefit from fewer attacks, as in Müllerian mimicry.

- **aggressive mimicry**
Aggressive mimics seek to deter attack by mimicking dangerous creatures of a different kind; a good example would be the elephant hawkmoth and its snake's-head startle pose.

Like Bates, Poulton regarded mimicry as actually an important section of special resemblance. Both Batesian and Müllerian mimicry are protective kinds. We need to remember that Müllerian mimics, besides copying each other, are showing warning coloration. They are doing the reverse of hiding: they sing from the same songsheet that they are not worth attacking. What unites all of these categories is the use of colour patterns to conceal, to obscure identity or intention, or to advertise dangerous qualities.

Poulton explained one of the most significant aspects of mimicry: its predominance in the insect world and why this has come about:

> The defenceless character of the group as a whole, the extent to which they are preyed upon by the higher animals, their enormous fertility, and the rapidity with which the generations succeed each other, are reasons why natural selection operates more quickly and more perfectly than in other animals, producing mimetic resemblances or other forms of Protective Resemblance in number and fidelity of detail unequalled throughout organic nature.

This is reminiscent of Darwin's magisterial passage in *On the Origin of Species* on the struggle for life, quoted in Chapter 1 (see p. 19). Poulton might have added: 'especially in the tropics', where the lack of seasons allows for a constant procession of the generations. Poulton's argument from the great fertility of insects explains the alleged 'excessive' perfection of the bird-dropping spider or *Kallima*. Out of this churning mill of reproduction and predation have come miraculous likenesses.

Thomas Belt, author of *The Naturalist in Nicaragua* (1874), gave a similarly vivid illustration of the tit-for-tat struggle between predators and prey which produces a gradual refinement of mimicry and camouflage:

> . . . natural selection not only tends to pick out and preserve the forms that have protective resemblances, but to increase the perceptions of the predatory species of insects and birds, so that there is a continual progression towards a perfectly mimetic form . . . Suppose a number of not very swift hares and a number of slow-running dogs were placed on an island where there was plenty of food for the hares but none for the dogs, except the hares they could catch; the slowest of the hares would be first killed, the swifter preserved. Then the slowest-running dogs would suffer, and, having less food than the fleeter ones, would

have least chance of living, and the swiftest dogs would be preserved; thus the fleetness of both dogs and hares would be gradually but surely perfected by natural selection, until the greatest speed was reached that it was possible for them to attain.

Poulton established many ongoing themes of mimicry research, especially concerning what mimicry can tell us as to whether the concept of the species rests on biological fact or is merely a convenient system of human classification. In mimicry, we say that one species is copying another; but this would be shaky if the species did not have a consistent and reliable identity. For Poulton, a species was a 'formed reproductive community', that is, an assembly of creatures which are normally found to mate with each other and with no other creature. He developed his ideas on speciation in response to a bound collection of the mimicry papers of Bates, Wallace and Trimen – a volume which Wallace gave him in 1903. Darwin believed that 'The only distinction between species and well-marked varieties is, that the latter are known, or believed, to be connected at the present day by intermediate gradations, whereas species were formerly thus connected'. This is, in fact, a good answer to the question people sometimes ask: 'Why can't we see evolution going on today?' We can and Poulton described what it looked like. For him, these races were, 'as it were, trembling on the edge of disruption, ever ready, by the development of pronounced preferential mating or by the accumulated incidental effects of isolation prolonged beyond a certain point, to break up into distinct and separate species'.

This is the situation today among the *Heliconius* butterflies. If Poulton made species seem less than solid in their outlines than we had thought, he also stressed that, despite this situation, species have a reality not possessed by such categories as 'genera', 'families' and 'orders'. The latter are ways of *classifying* relationships. Species are real in the way in which days and years are real, whereas the hour and the week are arbitrary divisions of time. Species are live, hot, breeding creatures, not classificatory pigeonholes. Their forms will drift over time (as does the length of days and years) so that the community of species at the present will differ somewhat from a community of the 'same' species at near or distant points in the past, but it still has a reality denied to the merely classificatory labels.

Although Poulton opened many eyes to the wonders of mimicry, he also contributed greatly to the scepticism of some critics. Because mimicry had been seized upon by Darwin as evidence for natural selection, it was always in the frontline for attack.

By a curious twist, the creatures that wished not to be noticed received so much attention from the mimicry enthusiasts among the entomologists that mimicry was fancifully sought out in everything: a 'bush was supposed a bear', or a bear a bush or an insect a leaf. So Poulton sometimes saw 'sermons in stones' where there weren't any. His most persistent critic was the American anti-Darwinian ornithologist Waldo Lee McAtee (1883–1962). McAtee was a down-home and dogged Mid-Westerner from Indiana, a 'Hoosier'* who mistrusted the academic zoological establishment and waged a lifelong war against Darwinian natural selection. A *causus belli* was the caterpillar of the lobster moth (*Stauropus fagi*), a favourite of Poulton's. This is an intriguing creature. Its food plant is beech; when at rest, the caterpillar resembles a withered beech leaf, but when it is disturbed, according to Poulton, it 'holds the anterior part erect, and assumes a terrifying position which mimics that of a large spider'. The caterpillar certainly strikes an attitude, but of what? The common name, 'lobster moth', comes from the caterpillar's trivial resemblance to an aquatic creature not often found on beech trees. Others have fancied that they saw in it something of an ant or earwig. So we can choose from beech, spider, ant, earwig, or lobster. To cap it all, there are black patches on the fourth and fifth segments which, Poulton claimed, were a deterrent to ichneumon flies, the marks resembling the results of a sting and 'the deceptive appearance of the traces left by an enemy suggesting that the larva is already "occupied" '. Poulton claimed that the lobster moth caterpillar needs all these defences because it is much persecuted by predators, but McAtee sarcastically observed:

Thus the larva of *Stauropus* is supposed to mimic more or less closely objects in both the vegetable and animal kingdoms . . . representatives of five orders . . . It is evident that the predaceous foes of *Stauropus*,

* A Hoosier is a person from Indiana; there are almost as many theories for the origin of the name as there are people in the state. A person who owns to the title is likely to be Mid-Western in outlook.

had they only the imaginative powers of its human observers, could have a banquet of many diverse courses, each of which would be merely *Stauropus* in disguise.

The kind of spat Poulton and McAtee engaged in was to be repeated by countless pairs of biologists until the present day. Much of the story of mimicry and camouflage revolves around such disputes. To anyone who is not a scientist, biologists' feuds seem deeply unseemly. Is it to this that the disinterested search for the truth about nature leads? Sometimes the quarrel has been between the wholly right and the wholly wrong, but more often a researcher has staked his or her life's reputation on a partial truth and tried to make this little truth cover all possible cases – to be *the* biological truth.

This is hubris. Biology is such a tangled web, it is a wonder that any generalising principles have been discovered at all. 'Natural selection' sounds like a binding law of nature: the equivalent of gravity in the inanimate realm: an omnipresent, inescapable force, which presses on every creature at every instant of its life. But life is not like that. The force of gravity acts in the same way on the huge mass of Mount Everest as on a paper clip, whereas natural selection impinges on the antelope and *Heliconius* differently: 'The race is not [always] to the quick nor the battle to the strong', but sometimes to the gaudy butterfly advertising the fact that it is inedible. Biology is the science of such exceptions in a way that physics is not.

CHAPTER 4

PANGENESIS

To solve the problem of the forms of living things is the aim with which the naturalist of to-day comes to his work. How have living things become what they are, and what are the laws which govern their forms?

William Bateson, *Materials for the Study of Variation* (1894)

As the nineteenth century turned into the twentieth, the mood in all of the arts and sciences changed. In physics the three-century rule of Newtonian mechanics was broken by the discovery of the quantum (the discontinuous nature of radiation) and by Einstein's relativity; painting broke free from naturalism in violent movements which liberated colour and form – fauvism and cubism. Biology was no exception. It began to go beyond the beguiling appearances, into the deep structure of living matter. Darwin had started out as a naturalist, an observer, but he had tried very hard to become an experimental scientist. The problem for him was that there were no ground rules on which to build: he was working in the dark.

To understand living matter, biology needed its atomic theory: some irreducible building blocks with which to comprehend the complexity life has developed. Without it, the study of mimicry would never get much beyond the naturalists' 'curiously exaggerated delight' in visual deception. In fact the twin pillars of such a basis might have been available during Darwin's lifetime, but circumstances dictated otherwise. Just how far Darwin's insight went beyond what could then be observed in nature has been highlighted by the contemporary biologist Sean Carroll:

> Darwin was asking his readers, in essence, to imagine how slight variations (whose basis was unknown and invisible) would be selected for (which occurred by a process that was also invisible and not measurable) and accumulate over a period of time that was beyond human experience.

Not only was the basis of the variations that natural selection worked on unknown and invisible; in Darwin's time, almost nothing was known about the nature of the things that varied. The question is: what are living things made of, how do they stay alive, and how do they reproduce. And is there a basic building block at the root of all living things?

There is, and it was known to Darwin: it is called the cell. When the microscope was invented in the seventeenth century, Robert Hooke had seen tiny cavities in cork and other plant material. In 1839, while Darwin was beginning to compile his evidence for evolution by natural selection, two German biologists formulated the cell theory: all living things are made of cells, small sacs, typically one hundredth of a millimetre across, with an outer membrane and mysterious contents which must bear all the secrets of life.

In 1859, the same year that Darwin published *On the Origin of Species*, a German pathologist and biologist, Rudolf Virchow (1821–1912), formulated a theory almost as dramatic as Darwin's for the future of biology: he claimed that every living cell originated from another living cell – *omnis cellula e cellula*. Darwin had been very much taken with the cell theory and realised that the idea of *omnis cellula e cellula*, coupled with the principle of evolution, meant that the whole of creation derived from the ability of cells to divide and multiply – to make a creature from an egg – and then for the result of this process to be subjected to variation and natural selection. He wrestled with these ideas in his 1868 book *The Variation of Animals and Plants under Domestication*.

But the behaviour of cells was very poorly understood. In trying to understand their role in inheritance, Darwin surmised that, for complex creatures to reproduce, every cell must throw off a germ cell which can be the seed to recreate the creature in the next generation. He called his theory 'pangenesis' (from which the word 'gene' was eventually derived) and the germ particles he called 'gemmules'.

Poor Darwin could never have guessed that his gemmules lived *inside* every cell and that it wasn't necessary for every single cell to throw off a germ cell, because each cell of a creature contains all the genes for the whole organism. But Darwin asked the right questions, as ever. He realised that the puzzle was this: how do dividing cells manage to maintain their identity even in different surroundings, such as when tissue from one animal is grafted onto another?

And what was the cell and what did it contain? Towards the end of Darwin's life something of what happens when one cell becomes two was observed by a German biologist, Walther Flemming (1843–1905). In 1878, he saw spindly structures that only appeared when a cell was about to divide: these structures were seen to pull apart and then a new cell wall formed, cutting the cell into two. Darwin had been looking for the particles that passed on the instructions for inheritance. Flemming's work suggested that they resided in this spindle, which appeared in the centre of the cell when it divided. Flemming called this active centre of the cell the nucleus. The spindles which Flemming had seen were, of course, the chromosomes.

What Darwin sought might have been discovered if Flemming's insight had been combined with some experiments on inheritance conducted by a Moravian monk, Gregor Mendel, in the early 1860s. Mendel showed that there were some simple rules for inheritance, with some characteristics being passed on without blending from the parent to the offspring (Fig. 4.1). To combine Flemming's and Mendel's insights would have given Darwin the essential clue to the nature of the living stuff which reproduced itself with surprisingly constant characteristics, but also with some variations. Yet these connections were made by no one until eighteen years after Darwin's death.

In 1900, in a perfect example of the old scientific folk-saw about waiting an eternity for a neglected theory to be rediscovered, three scientists – the Dutchman Hugo de Vries (1848–1935), the German Carl Correns (1864–1933) and the Austrian Erich von Tschermak (1871–1962) – independently announced to the world Mendel's

Crossing produces all black butterflies because every individual will receive one
black gene and one yellow, black being dominant to yellow. But the hybrids still have
the hidden yellow parental gene.

Hybrids are sterile but can be back crossed with the yellow parent.

The result is 50% black and 50% yellow butterflies.

Fig. 4.1   Mendelian inheritance demonstrated in swallowtail butterflies. Mendel explained
how traits can disappear when individuals are crossed but reappear in a future generation.
The example given here, of black and yellow forms of swallowtails, led to a medical breakthrough in the the 1960s (see Chapter 12). The phenomena of dominance means that a
pattern can be hidden by a dominant gene but resurface in a future generation.

findings, made thirty-four years earlier, in 1866. In that year, Gregor
Mendel (1822–84) had published a paper in the *Proceedings of the Brünn
Natural History Society* on the results of seven years of breeding experi-
ments with peas.

Mendel was not the simple self-taught friar of popular legend. Moravia
was then a province of the Austro-Hungarian Empire, and he studied

biology and physics under distinguished teachers (including Christian Doppler of Doppler Effect fame) at the University of Vienna. He was, nevertheless, an awkward character and failed his exams, not through lack of ability but from stubbornness. He eventually settled in the monastery at Brünn (now Brno), teaching and experimenting with pea plants.

There is a persistent legend that his paper, sent in the form of an uncut signature, lay unread at Down House for the rest of Darwin's life; but this cannot be substantiated. A very small group of people did read Mendel's paper, but it never entered the mainstream of knowledge until that rediscovery in 1900, long after Darwin's death. A book which did mention Mendel, W. O. Focke's *Die Pflanzen-Mischlinge* (1881), was acquired by Darwin only eighteenth months before he died. The legend of the uncut pages probably derives from pages 108–10 of this book, in which Mendel's work was mentioned.

Mendel's discovery of definite rules for inheritance was the most important biological theory since Darwin's. Mendel solved a problem that had plagued Darwin's theory and troubled the man himself. It was 'obvious' that offspring showed blended characteristics from their parents – there is a mixing and a diluting. Darwin believed this and was deeply troubled by its implications for his theory. Because, if all characters were blended in each generation, how could any new trait ever propagate itself over the millions of years of evolution? Each innovation would be a drop of coloured paint stirred into a can of white, and so drowned in a blanching sea.

Mendel had the answer to this. In the 1860s he set out to study the reproduction of peas, a system much simpler than many others. He selected pure-breeding strains of peas with different characteristics. There were seven paired traits in his experiments:

- seed wrinkled or round
- seed coat white or grey
- unripe pod green or yellow
- flower either terminal or axial
- height tall or dwarf
- flowers purple or white
- pod shape inflated or constricted.

He then began crossing plants with the opposed traits: wrinkled seeds with round ones, tall with dwarf, and so on. When he had offspring, he crossed them with each other to see which traits appeared when the first-generation peas were interbred.

He found that, in each paired trait, one was dominant (the modern term, which he introduced). So all of the offspring of a wrinkled/smooth cross were round. There was *no blending*. But the vital finding was that, when plants from this first hybrid generation were crossed, the wrinkled form *reappeared* in a quarter of the plants. This was how traits persisted: even if masked in one generation by a dominant factor, they might reappear in a future generation.

Mendel deduced the correct explanation for this. He believed that, when a wrinkled was crossed with a round, both traits entered the offspring but, round being dominant, the wrinkled form, though present in essence, was not expressed (such traits are called recessive). But when the hybrids were crossed, in a quarter of the crosses the offspring would have two copies of the wrinkled form; a quarter would have two copies of the round form; and half (from crosses both ways) would have one copy of the wrinkled, one of round, their appearance being round because of dominance.

Mendel's discovery was to play an essential role in the understanding of mimicry. Because mimetic patterns are inherited like Mendel's pea traits, where the peas have 'flowers purple or white', or 'seedcoat white or grey', the butterflies have 'forewing band red or yellow', 'hindwing band rayed or unrayed' and so on. Mendelian genetics was the first tool we had to understand these patterns.

Two other figures were important in bringing the organism back into biology in the late nineteenth century: Hugo de Vries, one of Mendel's co-rediscoverers, and William Bateson (1861–1926). If Mendel was lucky or perspicacious in his choice of a model organism, Hugo de Vries was unlucky. His experimental organism was the evening primrose (*Oenothera lamarckiana*), a North American plant which had been transported to Europe, rapidly colonising waste places. In 1886 de Vries, a zoology professor at Amsterdam, observed that new forms of *Oenothera* appeared quite often and bred true from the start. Around these observations he erected what he called the mutation theory.

This can sound misleading to the modern reader because, when we hear 'mutation', we think of genes; but genes were just a gleam in de Vries's eye at the time. De Vries believed that Mendel's ratios could be explained in terms of particles of inheritance inside the germ cells – he called them 'pangenes', from Darwin's 'pangenesis' theory (a name later on abbreviated to the modern 'gene'). These were, in fact, the 'gemmules' that Darwin had sought. But when de Vries talked of mutations he really meant the whole organism, which had jumped from one species to another – the idea of a mutation in the genes was still to come. This notion of a sudden jump from one species to a new one was contrary to Darwin's view that evolution proceeded by means of cumulative small variations.

De Vries was unlucky because the evening primrose is not very typical. It is now known that its new forms result from gene doubling rather than from mutation, as the latter is now understood. In other words, every cell of the new species contains an extra copy of all the genes, and this has some effect on the character of the plant.

Like de Vries, the Cambridge zoologist William Bateson did not believe that evolution proceeded by means of cumulative small variations. In his massive tome *Materials for the Study of Variation* (1894), he documented many monstrosities in all kinds of animals. Many of Bateson's monsters showed the substitution of one body part for another, for instance a fly's antenna positioned where its legs should be. He called this process of substitution *homeosis*, a name that is still in use today. It was clear to all that Bateson's monsters were produced in one leap, but the problem with suggesting this as an evolutionary mechanism was obvious. Such monsters were not viable. They were defective and often sterile.

But Bateson was onto something: if a human baby could be born with two hands joined at the thumb, this clearly tells us something about the *normal* process of making a hand, i.e. that there is a genetic instruction which commands a whole set of subroutines for making fingers and the other parts of the hand: once the gene has flipped into duplicating the 'hand' instruction, the rest of the program runs automatically to fill in the details. This is to interpret the phenomenon in the modern way – Bateson was not able to understand it like this. Nevertheless, Bateson's work was the first clue to nature's pattern-making powers.

Bateson's ideas entered the field of mimicry through the figure of Reginald Punnett (1875–1967), Bateson's assistant at Cambridge from 1902. He came from a well-known family of Sussex fruit growers, so dominant in their time that the standard container for holding fruit is still known as a punnett. In 1905/6, Bateson and Punnett discovered linkage, the biggest breakthrough in genetics after Mendel. Mendel had been fortunate in choosing characters of his peas which produced clean results, with simple ratios such as 3:1. We now know that, although such clean segregation is often the case, just as often it is not. Punnett and Bateson were working with sweet peas rather than with Mendel's garden peas, and they discovered that in two pairs of characters – purple-flower-dominant-to-white-flower and long-pollen-grain-dominant-to-round-pollen-grain – the characters did not segregate in the Mendelian way. The trait *purple flower* seemed to be linked to *long pollen grain* and *white flower* to *round pollen grain*. There was, however, a further complication in that a small number of plants were found with purple flowers and round pollen grains and vice versa – which suggested that the linkage could sometimes break down. The explanation for these complications was to come from an insect, not a plant, and from across the water.

Enter the fruit fly – *Drosophila*. Their wing patterns are rather boring, but fruit flies have much else in common with those other winged insects, butterflies. Throughout the twentieth century and into the twenty-first, the fruit fly has been in the vanguard of genetics. Why? Because it breeds very rapidly: fruit flies emerge from the eggs in just seven to eleven days. To draw conclusions from genetic experiments, thousands of crosses have to be made and analysed, and when flies can be raised as quickly as this the task becomes much easier.

The American Thomas Hunt Morgan (1866–1945) was the most important geneticist after Mendel. Morgan came from a Southern aristocratic family on his father's side. After Mendel was rediscovered, Morgan used *Drosophila* as a model organism, showing that Mendel's particles of inheritance were genes located on the chromosome. He won the Nobel Prize in 1933 for this work.

Morgan began by looking for mutations in fruit flies. Unlike de Vries's evening primrose, the fruit fly was very reluctant to produce any. Then, one day in 1910, the normally red-eyed fruit flies produced a single white mutant. So dramatic an event was this, the story goes, that, when Morgan's third daughter was born soon after the emergence of the white-eyed fly, his wife's enquiry, 'Well, how is the white-eyed fly?' set Morgan off on such an excited monologue that he had to stop himself to ask, 'And how is the baby?'

The white-eyed fly was a breakthrough in more ways than one. Now, one hundred years later, genes and chromosomes are the stuff of everyday conversation, but it is easy to forget that, in the first decade of the twentieth century, no one knew what was the connection (if any) between them. Chromosomes were the spindle-like objects Flemming had seen through the microscope when cells divided. They went through some strange contortions, and the fact that they split to form a new set in the two new daughter cells suggested they had some role in inheritance. On the other hand, the fact that they melted back into the cell after division suggested that perhaps they had no permanent structure and identity at all.

Genes, on the other hand, were entities inferred from Mendelian breeding experiments. Because whole number ratios such as 1:1, 3:1 and 9:3:3:1 were found in breeding experiments, these whole numbers were taken to represent some notional particles. But it was noticed that the chromosomes seemed to behave in a parallel manner to the genes. Perhaps they were one and the same thing, or at least closely linked?

Morgan was at first sceptical about both genes and chromosomes, pointing out that the Mendelian concept at that time was entirely abstract and notional and used not a single finding from the science of biochemistry: the whole idea of the genetic particles had no connection with the processes of living tissue.

Then came the white-eyed fly. Crossing the white-eyed fly (which was male) with a red-eyed female produced red-eyed flies. So the white-eyed gene was recessive. But then mating the first-generation hybrids produced a three-to-one ratio of red and white flies. This was the classical Mendelian ratio. But all the white flies were male.

Morgan drew profound conclusions from this. He realised that the inheritance of the white-eyed fly was sex-linked. His deductions approximated the modern understanding of the sex chromosomes, whereby the female has two x chromosomes. Genes were carried on the chromosomes: the red-eyed gene in the fly was carried on the x chromosome, and the white-eyed fly resulted from the absence or disabling of the red gene. This accounted for the crossing experiments.

Morgan then interpreted Punnett and Bateson's experiments in order to show how maps could be made of the genes' arrangement on the chromosome. Bateson and Punnett had shown that linkage sometimes broke down and that, in a small percentage of cases, two genes normally inherited together could be inherited independently. Morgan suggested that the work of a Belgian cell biologist, Frans Alfons Janssens (1863–1924), who in 1909 observed an intertwining of the chromosomes in cell division, could explain how linkage broke down. In this intertwining, portions of the chromosome could be exchanged. Morgan suggested that the frequency of linkage breakdown reflected how far apart the genes were on the chromosomes. In this way Morgan was able to produce 'maps' of the genes strung out along the chromosomes. The distance between the genes on the map was only relative – until the DNA era the actual nature and position of the genes were not known – but these linkage maps have proved their worth in locating genes ever since.

The linkage maps confirmed the chromosome theory: by 1914 Morgan had about two dozen mutants and they appeared in four different linkage groups. The fly, correspondingly, has four chromosomes; this could not be, and was not, a coincidence. Each linkage group represented a chromosome. The genes really were carried by the chromosomes. The modern era of genetics had begun. All of Morgan's hypotheses have been confirmed thousands of times over. Morgan's Fly Room at Columbia University, New York, was one of the cradles of modern biology and soon Morgan had mapped 100 genes across all the *Drosophila* chromosomes.

As far as butterflies were concerned, linkage maps were well in the future. The genetics of mimetic patterns was still a matter of conjecture, but on the question of how a mimetic resemblance might begin Reginald Punnett was prescient, noting that changes in the extent of black pigment are often employed in establishing a mimetic pattern:

A definite small change in the composition of the pigment laid down in the scales would result in the establishing of a mimetic likeness where there would otherwise be not even a suggestion of it. It is in accordance with what we know to-day of variation that such a change should appear suddenly, complete from the start.

These 'scales' need explaining. Everyone knows the dust that drops, especially from moths' wings, when they are handled. This dust, in fact, consists of tiny, intricately patterned scales made of chitin, a horn-like material. These wing scales, not possessed by other flying insects, are one of the defining traits of butterflies and moths. The wing patterns are composed entirely of these scales. As the insect develops from the pupa, the scales are squeezed out, like a kind of pasta, from the cells which make up the wings. Each scale comes from a single cell. Seen through the microscope, the analogy holds because, within the wing scale, under the surface, a structure of caverns and ribs is revealed – something like a pile of dry *festine*. But the scales are so small that, in our hands, they are mere dust. You can think of a butterfly wing scale as something like the nails of your fingers and toes. They are not themselves living but made of a tough, plastic-like substance squeezed out of living cells. The difference between nails and butterfly wing scales is that your nails are the composite product of billions of cells, whereas each butterfly wing scale is one single cell.

The scales are crucial to understanding butterfly wing patterns because they function rather like computer-screen pixels. Each scale only ever has one colour, so the complex coloured patterns are created by using different coloured scales, as in a micro-mosaic.

Punnett's conclusion was that natural selection and developmental mechanisms conspired together to create mimicry. This was seen as highly heretical for about a hundred years, but it does not seem so outrageous now: 'On this view natural selection is a real factor in connection with mimicry, but its function is to conserve and render preponderant an already existing likeness, not to build up that likeness through the accumulation of small variations.' This view inevitably brought Punnett into conflict with the Darwinians, and the most notable Darwinian exponent of mimicry, Edward Poulton, took up the cudgels. To some extent, Poulton and Punnett represented different cultures: both were upper middle class, of course, but Punnett was a Cambridge man and a keen cricket player, Poulton was an Oxford man and inclined to write those light verse ditties. Their quarrel revolved especially around the swallow-tail butterflies.

As Wallace knew way back in 1854, a single brood of eggs from a mimicking swallowtail could produce four different kinds of butterfly: non-mimicking males and three different females, all of which mimicked other (toxic) species. No intermediate forms were ever discovered. As a devoted Mendelian, Punnett speculated that, since there seems to be one gene controlling all the patterns – a phenomenon known as polymorphism – the mutations to create the new wing patterns must have emerged suddenly and in a complete form.

If the patterns had been laboriously compiled, like a jigsaw, through small cumulative variations over immense periods of time, on the Darwinian principle, the pieces of the jigsaw would be controlled by different genes, and these would separate out in crosses. So for Punnett, the fact that the patterns are inherited entire, either one or the other, meant that the mutation which produced the new form must also have been a one-off, all-or-nothing event.

Every researcher is influenced by his or her particular area of expertise. Punnett's best genetic work was done with sweet peas and they, like many plants, do produce 'sports' – sudden dramatic mutations. He cited as evidence a new dwarf sweet pea variety called 'Cupid'. So why shouldn't the swallowtail harem obey the same rules as the 'Cupid' sweet pea? Should not love among the butterflies be the same as among the sweet peas? But Poulton was sure that this was nonsense: even if a swallowtail had once had a freak mutation, one which produced a like-ness to a toxic model species, this could not have happened three times.

Being English gentlemen, Punnett and Poulton resolved in 1909 to travel together to Ceylon (now Sri Lanka) to settle the matter; but in the end Poulton could not travel, so Punnett went to study the swallowtails there without him. His visit led to detailed crossing experiments which confirmed that the swallowtails always produced perfect mimics with no intermediates.

Their tussle was taken up again in a short-lived journal of biological controversy, *Bedrock*, in 1913–14. Punnett managed to extract from Poulton an admission that 'I have always recognised that the first varia-tion [in forming mimicry] must be something appreciable, something which, at any rate, at a distance and on the wing would recall the pattern of the model.' This became the favoured compromise position: mimicry must have started with a new pattern which resembled sufficiently that of a toxic model species; then natural selection could get to work.

The swallowtails really were odd. With the best will in the world, it was hard to understand what was going on. Punnett tried to sum it all up in *Mimicry in Butterflies* (1915), the first book devoted entirely to the subject. Here he began the serious process of enquiring by what mechanism mimetic species acquired their patterns. He noted that mimics and their models ran in series, with many model and mimic species in a few families. For Punnett, this meant that there was a limited range of pattern-making machinery in butterflies, and so it was relatively easy for supposedly unrelated butterflies to hit upon the same patterns.

Punnett drew an analogy from animal coloration. There is a series of coat colours common to rabbits, mice and guinea pigs: agouti, black, chocolate, blue-agouti, blue and fawn. These animals are more different in appearance than one butterfly is from another but, deep in nature, there seems to be a biological paint-colour chart which is accessible to many different species.

Punnett noted that many butterfly genera are not mimetic, presumably because the developmental mechanisms in such genera do not facilitate the right kind of pattern formation. But, when the mechanisms are propitious, mimicry is quite easy to achieve, because the creatures are shuffling that limited pack of patterns. This is an oversimplification, but at the time it was a powerful insight.

Punnett and Poulton approached the same problem from different directions. Although Darwin had done his best to understand inheritance and the workings of the cell, his followers seemed nonchalant about the actual composition of the living matter on which that natural selection acted. Geneticists like Punnett and Bateson, working with the actual living creatures, breeding them and observing the offspring, were sure something was going on beyond that remorseless mill of abstract minor variations. In a sense, both camps were right, but in 1915 no resolution was in sight.

H.G. Wells might have had Poulton and Punnett in mind when he wrote of the fictional dispute between Professors Hapley and Pawkins in his story 'The Moth'. Biology is fertile territory for turf wars, as we have already seen with Poulton and McAtee. Hapley and Pawkins are taxonomists, experts in classification: a notoriously hair-splitting discipline, perfectly constituted to allow the bitterest disputes over the finest of distinctions. Pawkins dies and thereby removes Hapley's real reason for existence, the pursuit of the feud: 'For twenty years he had worked hard, sometimes far into the night, and seven days a week, with microscope, scalpel, collecting-net, and pen, and almost entirely with reference to Pawkins.' Pawkins begins to haunt Hapley in the form of a ghost moth:

> Once he saw it quite distinctly, with its wings flattened out, upon the old stone wall that runs along the west edge of the park, but going up to it he found it was only two lumps of grey and yellow lichen. 'This', said Hapley, 'is the reverse of mimicry. Instead of a butterfly looking like a stone, here is a stone looking like a butterfly!'

Between them, Punnett and Poulton posed many of the questions about mimicry that are still wrestled with today. That their approaches seemed incompatible only highlighted the difficulty of the subject. Perhaps the truth would reconcile their different positions rather than validate one and falsify the other? But the subject was not destined to remain the sole province of naturalists and biologists. Visual resemblance obviously interests artists, and in the 1890s Poulton had encountered a painter with something to say about mimicry and camouflage.

CHAPTER 5

ON THE WINGS OF ANGELS

Only an artist, perhaps, can rightly appreciate the profound and perfect realism of these background pictures woven by birds and other animals.

Abbot H. Thayer, *Concealing Coloration in the Animal Kingdom* (1909)

In the 1890s, an eccentric, highly opinionated New England painter gate-crashed the small circle of naturalists concerned with mimicry and camouflage. Abbott Handerson Thayer (1849–1921) – one of the few artists to have a scientific law named after him (Thayer's Law of Concealing Coloration) – is an intriguing study in art, science and the human ego. The 'two cultures' divide went straight through him and caused him much distress – although, ultimately, this was more a product of his strange temperament than of any schism between art and science.

Although his ideas were accepted by many zoologists and Poulton became his friend and champion, Thayer's views were extreme. He believed that all animal patterns and coloration had concealment as an end and that even brightly coloured animals apparently exhibiting warning coloration and mimicry were in fact camouflaged. His dogmatism on this point led to many heated and often comical confrontations.

Thayer was in some respects a typical New Englander at a time when New England evinced an idealist, ruralist ethos: Thayer's favourite authors were Emerson, Robert Louis Stevenson and Mark Twain. The son of a country doctor, he was born in Boston and at seven moved to

New Hampshire, where he quickly became immersed in nature, learning to hunt and trap and devouring Audubon's *Birds of America*. He discovered painting early and by the age of eighteen had set himself up as a painter in Brooklyn.

Besides nature, Thayer's principal subject was women, about whom he had some strange ideas. Thayer was an intense realist in his portrayal of nature; but he also extolled, in Emerson's words, 'the noble woman with all that is serene, oracular and beautiful in her soul'. In his life as well as in his painting, women were all important. In 1872 he married Kate Bloede, a German girl whose Teutonic romanticism reinforced his New England idealism. Thayer surrounded himself with female students and female assistants, many of whom he painted.

From 1875–9, Thayer and his wife lived in Paris, where he studied. But Thayer remained aloof from the progressive tendencies in French art. In 1887 his art took a strange turn with a portrait of his daughter, Mary: the women in his paintings, otherwise realistically portrayed, began to sprout angel's wings. Thayer gave his reasons:

> Doubtless my lifelong passion for birds has helped to incline me to work wings into my pictures: but primarily I have put on wings probably more to symbolize an exalted atmosphere . . . when Mary poses she fits into them and completes the picture . . . you . . . can believe what a beautiful sight it makes . . .

From 1888, the Thayer menage lived during the summer in a house built for them at Dublin, New Hampshire, by one of his assistants. Here animals had the run of the range and Thayer lived the life of a New England backwoodsman – a type familiar at the time. Thayer's biographer, Nelson C. White, wrote:

> Thayer was much like Thoreau, who took fish from the water in his hands, pulled the woodchuck by his tail from the burrow, and lifted the partridge from her nest. Sometimes his familiarity and nonchalance with wild animals was disconcerting, not only to casual and unsophisticated callers, but even to the family itself, as when he put a porcupine on his shoulder and it made itself at home by playfully biting his ear.

From 1892 on Thayer began to observe the way in which many animals were darkly coloured on the back and pale on the belly. His original contribution to biology was to explain the reason for this effect. When strong light falls on the back of an animal, whatever colour it is, the effect is of a white glare, and the underside, correspondingly, is in shadow, appearing darker than its natural hue. So animals have evolved a compensating colour scheme which reduces the contrast in lighting and makes them less distinguishable from their background, lightening the colour where the creature is habitually shaded – the underside – and darkening the back where the sun's glare bleaches out colour. As Thayer put it: 'Animals are painted by Nature, darkest on those parts which tend to be most lighted by the sky's light, and *vice versa*.' This animal colour scheme is known as countershading.

Thayer's law refers to one aspect of natural camouflage only: the gradation between the back and the belly of an animal. So rabbits, hares, rats and mice are brown on the back grading to white on the belly. Many fishes show a similar pattern. Thayer commented that he had discovered the principle through the difficulty he experienced in painting animals in their natural habitat. The painter's approach is, in some respects, the antithesis of nature's. To create the illusion of a solid form on paper or canvas the painter resorts to modelling, that is, using gradations of colour to create the appearance of light and shade; it is this that gives paintings their suggestion of three-dimensionality in our eyes.

The zoologist Hugh Cott (1900–87), the key figure in camouflage and mimicry in the mid-twentieth century, described the contrast between art and nature eloquently:

> The artist, by skilful use of light and shade, creates upon a flat surface the illusionary appearance of solidity: nature, on the other hand, by the precise use of countershading, creates upon a rounded surface the illusionary appearance of flatness. The one makes something unreal recognizable: the other makes something real unrecognisable.

The painter has to create an impression of solidity and a strong outline, but Thayer found that wild creatures always blended into their background to some extent, which made them hard to pick out in painting their images. The outline was hard to see distinctly against the background and, even worse, where you expected shadows to fall the creature showed a lightening of coloration.

It is possible to exaggerate this flattening effect, and it seems that Thayer was abnormally sensitive to it. Most people, observing a rabbit or a deer or a thrush, will notice, if it is drawn to their attention, that the underside is much paler than the back; but that the creature looks 'flat' probably would not occur to them. It did to Thayer, and it became his obsession. In 1896, he published his findings under the title 'The law which underlies protective coloration' in the natural history journal *The Auk.*

Thayer's ideas on countershading were welcomed by most biologists; indeed, Poulton had more or less discovered the same phenomenon in 1886, but he recognised that Thayer had gone deeper into the subject and he became his champion, presenting Thayer's theories to a British audience in *Nature* in 1902.

Prior to this, Thayer visited England in 1898, where he demonstrated his principles:

> I have set up a pair of models of birds (cork, painted) in a glass case, one at Oxford and one at Cambridge . . . They are on an axis with a crank and the visible one is painted all over with the same paint as the background and the other so well graded that the delighted zoologists can't see it at all at five or six yards.

Thayer was not a scientist; he had nothing of the scientist's temperament. He was an artist whose idealist fervour, edged by deep insecurity, led him to regard his findings less as discovery than as revelation. His quasi-religious feelings about artistic vision soon induced delusions of grandeur. Having discovered a principle of the natural world – countershading – and being a painter of lofty vision, Thayer felt able, in the words of Ross Anderson, curator of a 1982 retrospective of Thayer's work at the Everson Museum, Syracuse, to identify 'God as a professional colleague (albeit a superior one)'.

By the time his ideas had been consolidated in the magnum opus he co-wrote with his son Gerald – *Concealing Coloration in the Animal Kingdom* (1909) – Thayer's prophetic intolerance was in full flood. He boasted, as no scientist ever would: 'Our book presents not theories, but revelations, as palpable and indisputable as radium X-rays [sic].'

Thayer's comical over-assertiveness was really over-compensation. From an early age, he had a deep need of approval for his paintings and he sensed opposition even when there was none. When it came to his work in zoology, he did not understand that scientific ideas are not accepted on the basis of the vehemence with which they are expressed. Thayer was exceptionally proud of having discovered a scientific law and proud that it was an artist who had discovered it. He often remarked intemperately: 'An *artist* is of course the judge of such copies; and it is therefore as an expert that I pronounce on them.' The battle for the right to the keys of the kingdom of camouflage between artists and biologists begins with Thayer:

> The entire matter has been in the hands of the wrong custodians. Appertaining solely to animals, it has been considered part of the zoo-logist's province. But it properly belongs to the realm of pictorial art, and can be interpreted only by painters. For it deals wholly in optical illusion, this is the very gist of a painter's life.

Why, one asks, '*only* by painters' – might not scientists at least be able to learn some of these interpretations?

Being a painter, Thayer illustrated the thesis with his own water-colours. He posed animals against backgrounds which purported to show that at perhaps one instant of their life they were perfectly concealed, even if most of the time they stood out a mile.

Most people take the male peacock to be an extreme example of sexual display. Indeed, its magnificent brocaded fan seems a liability in survival terms, not only in its gaudy colour but in its burdensome weight and unaerodynamic profile. But Thayer believed the male peacock to be exquisitely camouflaged; he painted a watercolour of one in a forest glade to demonstrate the point. Thayer's caption states:

> The peacock's splendour is the effect of a marvellous combination of 'obliterative' designs in forest-colors and patterns. From the golden– green of the forest's sunlight, through all its tints of violet-glossed leaves

in shadow, and its coppery glimpses of sunlit bark or earth, all imaginable forest-tones are to be found in this bird's costume; and they 'melt' him into the scene to a degree past all human analysis.

At which point we wonder: if the male peacock is so well camouflaged, presumably the dowdy peahen is not? Most observers would consider the reverse to be the case.

The most bizarre example painted by Thayer, and one that was to earn him eternal ridicule, was his illustration of flamingos feeding at a lake and 'camouflaged' against the sunset. For Thayer,

> These traditionally 'showy' birds are, at their most critical moments, perfectly 'obliterated' by their coloration. Conspicuous in most cases, when looked at from above, as man is apt to see them, they are wonderfully fitted for 'vanishment' against the flushed, rich-coloured skies of early morning and evening.

Flamingos feed by burying their necks in shallow salt lakes, but there is no reason for them to disappear against the red sky. Thayer's friend, the artist Royal Cortissoz, had a convincing explanation for Thayer's bizarre championing of the flamingo as a prime example of camouflage: when he observed them feeding in the West Indies, Thayer was 'watching more than a fact – he was watching a "big magic", and the artist in him was thrilled'.

Reading Thayer's book today is a strange experience. He sets out with the idea that *every single creature* is perfectly camouflaged, and then tries to show how this works in practice. He has many photographs with captions asserting that an animal has 'disappeared' when the photograph plainly shows the opposite. It is as if Thayer practised auto-suggestion: he has convinced himself of this perfect camouflage and then tries to bludgeon his readers into doing the same.

Because, for Thayer, everything is camouflaged, nothing is allowed to exhibit warning coloration. Creatures that scream their warning signals, such as skunks and wasps, tempt him into the most unlikely assertions. Wasps are 'fundamentally and very potently obliterative against their wearers' average backgrounds of green vegetation in sunlight and shadow, and also amidst yellow flowers'. But every child (and every other animal) learns to fear their yellow and black stripes as a warning of their sting. Wasps might, in some dappled sunlight scenes, blend into the background; but this is not what the stripes are for.

Thayer was thoroughly Darwinian in spirit, but his version of natural selection had only one strategy:

> All patterns and colors whatsoever of all animals that ever preyed or are preyed on are under certain normal circumstances obliterative . . . Not one 'mimicry' mark, nor one 'warning color' . . . nor any 'sexually selected' color, exists anywhere in the world where there is not every reason to believe it the very best conceivable device for the *concealment* of its wearer . . . I believe it [countershading] will ultimately be recognised as the most wonderful form of Darwin's great law.

In a footnote, Thayer grudgingly admits that 'Animals' markings doubtless serve in various lesser degrees most of the purposes that have been attributed to them'; but he was simply not interested in the idea of mimicry. He devotes a passage to demonstrating that the mimetic Amazonian butterfly *Heliconius melpomene* is not warningly coloured but when at rest is camouflaged as a leaf. He asserts: 'The cases of out-and-out butterfly mimicry are relatively few indeed, and scattered, while "obliteration" is universally and most variously achieved in them.' So foreign to him was the notion of warning coloration that he imagined *Heliconius* butterflies to be defenceless: 'Probably this *Heliconius* finds such an intoxicating feast in the tops of certain great flowering trees that it becomes an easy prey to any birds that want it.'

Thayer's opposition to Batesian mimicry, in particular, was extreme. He and his family spent holidays in Bermuda, where Thayer revelled in the exotic wildlife. In 1903 he made an expedition to the island, accompanied by his family, with the express purpose of refuting Bates's theory of mimicry. His daughter Gladys recorded:

> My father's special mission was *tasting* butterflies! This was in order to disprove what his very dear friend, Professor Poulton of Oxford has written many lengthy books to *prove*, the theory of mimicry, trying to

show that harmless butterflies or other insects had through natural selection acquired similar patterning and coloring to those bad tasting butterflies for their protection. He actually tasted them and could find no difference in the flavour.

The question of the bad-tasting butterflies will recur throughout this book. For now, it is enough to say that the reactions of a human being to the taste of these butterflies – something natural selection never had to contend with – could hardly be the last word.

The problem with Thayer's extreme form of protectionism is that he believed that for a creature to blend in perfectly for only an instant might be life-saving, and hence might come under the sway of natural selection. But these freakish stopped-clock-is-right-twice-a-day coincidences would have to happen a statistically significant number of times for natural selection to come into play.

Thayer had an inordinate belief in the power of demonstration. If only people would *look*, he contended, they would be convinced. In the autumn of 1910 he staged one of these demonstrations at the Smithsonian Museum in Washington. More than looking, Thayer demanded participation. His friend, the artist Royal Cortissoz, commented on the scene: 'our principal and quite unforgetable exhibit was a young stuffed prongbuck, worn by Thayer like a yoke over his head and shoulders as we marched forth to meet the enemy'. Thayer asked the company to lie on the ground to see the prongbuck against the sky. They demurred.

Thayer's problem was that a good and true principle in nature does not have to be universal: biology is the science of exceptions. But, however much he exaggerated the scope of his ideas, a kernel of truth remained – the countershading principle survives and is still called 'Thayer's law'.

Thayer identified a second mechanism of camouflage: this was what he called 'ruptive coloration' and is now called 'disruptive coloration'. The rationale for using disruptive rather than concealing coloration is that there is a fundamental difference between large and small creatures. Insects especially can blend completely into their surroundings and can often enhance their invisibility by remaining motionless for long periods. But concealing the outline of large animals is very difficult. Another principle is needed here, and that is disruptive coloration. By breaking up the shape of the creature into large, seemingly random patches of colour, the characteristic outline of the creature can to some extent be obscured. As humans are large creatures, and their artefacts often larger still, this principle is more important in human camouflage than attempts at total invisibility.

Disruptive coloration is, in a sense, the opposite of countershading and of textures which blend in with the background. The idea is to 'paint' bold patterns onto a creature to break up its outline. There is no doubt that this technique does create intriguing new patterns out of the old form, and sometimes and in some creatures it is easy to see how it works. The Gaboon viper (*Bitis gabonica*) seems to have two jagged chunks cut out of its flattened head and a thick rope running down its back in sharp relief. On a leaf-strewn ground, the outline of the snake is easily lost.

The skunk, for Thayer, exhibits disruptive coloration, which he defined as 'bold, massed patterns of contrasting shades and colors, disposed at seeming haphazard over the animal's body, but in reality arranged according to the rigid laws of disguise'. So for Thayer the skunk is, in effect, camouflaged: 'Skunks . . . long believed by naturalists to be colored for warning conspicuousness (proclaimant of their foul defensive equipment), have, in fact, the universal obliterative coloration.' For him, 'warning conspicuousness' was just an obsolete theory, to be discarded in favour of Thayer's Universal Principle of Obliterative Coloration. In this interpretation, the white portions of a skunk are there to match the brightness of the sky when seen by small prey from below. And this skunk camouflage theory is 'attested beyond question by our photographs, as the reader will agree'. Anyone who has seen footage of a skunk displaying its aggressive posture, standing on its forelegs, fanning its tail, flashing its black-and-white body and ready to spray, is unlikely to be Thayer's compliant reader.

Thayer has many examples of seabirds with disruptive coloration. Guillemots, for instance, have a kind of yin-yang, black-top-white-belly interlinked pattern. For Thayer, 'seen against the sky (or brightly sky-lit sea) they "lose", so to speak, their light parts; seen against the shadowed rocks they "lose" their dark parts, and thus their bird-like contours are disguised'. The problem is that, although the pattern does split the bird into two and a single guillemot against rock or sky could easily be lost, guillemots mass in tens of thousands on the rocks of their breeding grounds, as his photograph shows. Such a pullulating mass of birds cannot 'melt into the background'.

Ideas as provocative as Thayer's were bound to find an adversary, and he became the *bête noire* of a distinguished American: none less than Theodore Roosevelt (1859–1919). Roosevelt, the twenty-sixth president of the United States, served from 1901 to 1909 and was an energetic, independent-minded and popular figure. On leaving the presidency in 1909, he devoted much of his time to his passion for big game-hunting and he published a book: *African Game Trails* (1910). Roosevelt was a big, bluff practical hunter who had little time for the theories of a painter of women with angels' wings. He devoted a twenty-page appendix in *African Game Trails* to attacking Thayer's ideas on camouflage. Following this, the two adversaries traded blows for some years in magazines and private correspondence. More heat than light was generated.

Roosevelt did not dispute that small creatures such as insects were often brilliantly camouflaged, but he doubted whether any such pattern was effective with larger animals: 'The theory is certainly pushed to preposterous extremes.' On the African savannahs there is nowhere to hide: 'No colour scheme whatever is of much avail to animals when they move, unless the movement is very slow and cautious.'

In their arguments, neither concedes an inch. Thayer believes that the two white patches on the rump of the prongbuck obliterate its outline. Roosevelt counters by saying: 'Ten steps farther back, or ten steps farther forward, would in each case make it visible instantly to the dullest-sighted wolf or cougar that ever killed game.' If Roosevelt asserts that the zebra is always highly visible to a predator such as a lion, Thayer counters that a lion looks up from three feet at a zebra's five, and this makes all the difference. If Roosevelt states that many of Thayer's so-called camouflaged creatures are only in harmony with the environment for a fraction of their lives, Thayer counters that a lion's teeth and claws are hardly ever used; most of the time the creatures lie around indolently – does this mean that these organs are not useful at the time of capturing and eating prey? The non sequitur in this argument – that the lion is sustained for several days by a single kill and clearly its teeth and claws are its way of life; a prey species is vulnerable at every instant when it is not miraculously in harmony with its environment, and its principal life strategy may not be camouflage – escaped Thayer. He was more intent on asserting himself at any cost than on reaching the truth. His defence is littered with blustering outbursts: 'It is the nearest to 100 per cent of error that I have ever read, on any subject that I understood'; 'Each of these facts is here to stay'.

One side of Thayer's work *did* earn Roosevelt's whole-hearted approval. In a footnote he says:

> In passing I wish to bear testimony to the admirable work done by various members of the Thayer family in preserving birds and wildlife – work so admirable that if those concerned in it will go on with it, they are entitled to believe anything in the world they wish about protective coloration.

Roosevelt was referring to Thayer's pioneering work in bird conservation. Around the turn of the century, bird feathers were so fashionable an adornment for women's hats (a trivial example of human/animal mimicry) that some species such as egrets and terns were threatened with extinction. Thayer campaigned for warden protection of their breeding sites. As such, he was one of the inspirations behind huge modern conservation organisations such as the Audubon Society in America and the Royal Society for the Protection of Birds (RSPB) in Britain.

Roosevelt was particularly scathing about zebras, which Thayer regarded as the acme of disruptive coloration, and indeed the zebra remains a contentious subject. The zebra's pattern is disruptive, but is it any use? Roosevelt noted that, to be effective at water holes with reedbeds (Thayer's favourite pose for the animal), the zebra's disruptive patterning would need to be motionless, and the animals are hardly ever in this situation.

Thayer was not the only artist to be fascinated by disruptive coloration. Cubism – a technique in which the conventional outline of a person or object is broken up by geometrical patterns – came to America in the form of the famous Armory Show of 1913. This exhibition, in New

York from 17 February to 15 March 1913, marked the sudden irruption of European modernism into the hitherto conservative world of American art.

At this point, European art had seen fifty years of innovation, mostly concerned with the process of painting itself: visual perception and conceptualisation rather than subject matter. Impressionism took naturalistic scenes and softened their focus, dissolving the outlines, heightening the colours of freakish weather effects to the detriment of normal coloration. Seurat experimented with the scientific theories of the chemist Michel Eugène Chevreul (1786–1889), producing colour harmonies through a multitude of tiny coloured dots rather than with mixed washes and strokes. Colour per se became more important and, in Van Gogh, assumed a hallucinatory vividness. False colour, flattened perspective, and a stylisation of outlines not seen in European art since the prehistoric cave painters appeared in Matisse, Derain and their followers around 1904–5. This group was known as the Fauves (wild beasts) on account of their uninhibited technique, throwing off artistic decorum in favour of garish colour and crude outlines. The element of stylisation was reinforced by the discovery of African art by artists such as Picasso and Matisse. Finally, pure abstraction began to appear in the work of the Russian painter Wassily Kandinsky around 1911.

The most important art movement from the point of view of camouflage was cubism, which developed after 1907, following Picasso's *Les Demoiselles d'Avignon*, with its distorted images of women. The tendency towards colour for colour's sake, so notable in many of these movements, was reversed in cubism. The palette was, more or less entirely, muddy greens and browns – earth colours, camouflage colours. Just as the Impressionists saw the visual world in terms of coalescing clouds of colour, the cubists saw it as geometrical facets – triangular, trapezoidal. Although cubist paintings always had recognisable subjects, the geometrisation involved sometimes came close to reducing them to abstraction.

The modernist movement, in music and literature especially, besides painting, was the greatest flowering of art since the Renaissance. But, for Thayer and some of his fellow American artists, it was an aberration, something to be ignored or kept at arms length. New York's emergence as the world capital of avant-garde painting was still forty years away.

With works by Matisse, Duchamp, Picabia, Picasso, Cézanne and Braque among others, the Armory Show had the same effect – galvanising young artists and being excoriated by philistines and the old – that Roger Fry's Post-Impressionist show* had on the London art world of 1910. The most controversial exhibit at the Armory show was Marcel Duchamp's *Nude Descending a Staircase*, likened by one critic, rather wittily, to 'an explosion in a shingle factory'.

Thayer and his friends were appalled. Royal Cortissoz was 'the bell-wether of the naysayers', castigating as 'foolish terrorists' those who wanted 'to turn the world upside down'. For Thayer, the distance between his pious 'angelisations' of virginal women and cubism's brutal deconstruction of reality into a multifaceted mask could not have been greater.

Thayer's consuming interest in the process of visual representation in painting – his technical knowledge of light and shade and of the tricks of perception – might have created a bridge to these exciting developments, but he set himself against them. There was a deep divide between the analyst of nature and the man who came to the canvas with angel-winged-woman in mind. But, if he had not been so emotionally repelled by cubism, Thayer might have noticed the connection between disruptive coloration and cubism's breaking up of the outline into facets. Cubism and disruptive coloration are obviously very different in some respects – cubism being a wilful fracturing of the two-dimensional picture plane for aesthetic reasons; disruptive coloration, nature's way of disguising the form of a three-dimensional creature – but in 1913 the two techniques were about to converge on the same end for the purposes of war.

Thayer was one among many artists and some zoologists who realised that the coming war would need techniques of visual deception. Thayer's disruptive coloration and the cubists' distortion of the normal two-dimensional effects of painting were two possible approaches. But, when cubism was thrust into his attention by the Armory Show, Thayer could see only the cubists' rejection of his high ideals for art and, for want of

---

* Fry's exhibition, *Manet and the Post-Impressionists*, included works by Cézanne, Gauguin, Van Gogh, Picasso and Matisse. The reactions were probably even more scandalised than for the Armory Show, Wilfrid Scawen Blunt huffing: 'Nothing but the gross puerility which scrawls indecencies on the walls of a privy'.

them, these artists merely showed 'an increasing lateral swing from one excess to the opposite'. When war came, Thayer showed a missionary zeal to instruct the military authorities in his principles of camouflage. That this aberrant art movement from Paris might also have a role to play would have seemed inconceivable to him.

CHAPTER 6

DAZZLE IN THE DOCK: THE FIRST WORLD WAR

For England to see these facts would treble her power before the end of a week; that is all the time it would take to paint every vertical inch of the whole navy, spars, cables and all pure white.

Abbott Handerson Thayer, 1916

One of the more genteel battlefronts in the two twentieth-century world wars was that between artists and naturalists for the keys to the kingdom of camouflage. Visual deception in combat begins in the natural world: the disguises of butterflies, spiders, mantises and other creatures are ploys in their struggle for survival. But the tricks of representation – creating a two-dimensional image of objects that produces the illusion of three-dimensional form – are the province of the artist. And when it came to human life-or-death war, both naturalists and artists had to convince hard-bitten military men that camouflage really could be effective.

Two men were the prime evangelists for camouflage in the First World War: Abbott H. Thayer, naturally, and the Scottish zoologist John Graham Kerr (1869–1957). At least this artist and this naturalist were in harmony, with Kerr espousing and promoting Thayer's principles. Thayer might have entertained a comically excessive belief in the power of natural camouflage, but his principles of countershading and disruptive patterning might well have relevance for human warfare. When war broke out between the USA and Spain in 1898, Thayer, as an expert on natural camouflage, was called in by the US Navy to suggest camouflage for ships. Nothing came of this at the time, but in 1902 Thayer took out a patent with Gerome Brush (1888–1954), the son of his close painter friend George de Forest Brush (1855–1941): 'Improvements in Process of Treating Ships and other Objects to Render them Less Visible'. The patent is a very simple document and seeks to apply to ships only the countershading principle. So, to counteract natural glare and shadow, upward-facing surfaces should be painted dark, vertical surfaces light, downward-facing surfaces very light, preferably white. Curved surfaces should grade from dark to light by imperceptible degrees; this is especially important with structures such as gun barrels.

Thayer believed that white was the key to camouflage at sea, an idea which was to prove highly controversial. His views became widely known when the *Titanic* sank on the night of 14 April 1912. Thayer asserted that, contrary to common belief, the white of a large iceberg is not conspicuous at night: on the contrary, at night – or even under a heavy cloud at day – white objects at sea are the hardest ones to see. This idea would be vigorously debated over the course of two world wars.

John Graham Kerr was professor of zoology at Glasgow University from 1902 to 1935, when he resigned to become MP for the Scottish universities. He had absorbed the principles of countershading and disruptive ('dazzle') coloration from Poulton and Thayer. He was also passionate about ships and, in 1895, while sailing through the newly opened Kiel Canal on the yacht *Raven,* he had observed French and German battleships in their standard grey livery and realised 'the extent to which man in his war camouflage falls short of what is attained by nature in the animal kingdom'.* This was a defining moment for Kerr and, when war came, he began to bombard the authorities with advice on camouflage, starting with Winston Churchill, then First Lord of the Admiralty. On 24 September 1914 Kerr sent Churchill a three-page outline of his 'method of diminishing the visibility of ships at a distance'. In his covering letter he rather optimistically recommends that Churchill visit the Museum of Zoology in Cambridge, where one of Thayer's model birds would 'enable you at once to appreciate the point about

---

* Kerr was writing in 1941.

gradual shading'. So began an intense struggle to find the best way to camouflage ships at sea.

Kerr's memorandum draws heavily on Thayer. It begins by dismissing the vulgar idea that merely copying the colour and tone of the background would confer invisibility on ships. But it advocates very strongly Thayer's two principles: countershading (here called by Kerr 'compensating shading') and disruptive patterning. Kerr doesn't dwell too long on animals; he makes the general case for camouflage in nature, then moves on to the abstract physical principle of countershading, and finally to ships. To obliterate the strong shadows on the upper parts of a ship, Kerr advocates picking out the deep shadows in brilliant white and grading smoothly in tone between these white-painted areas and the brightly lit areas painted in grey. Big guns should be grey on top shading to pure white beneath. All this is in line with standard Thayer theory.

Kerr then moves on to disruptive patterning. He cites the zebra and recommends that the outline of ships be broken up by patches of white. He makes much of the masts, which should have irregularly edged bands of white in order to break up their vertical line. The purpose of the mast camouflage was not merely to make the vessel less conspicuous but to 'greatly increase the difficulty of accurate range finding'. The paper concludes with a strong, reiterated recommendation: '*to destroy completely the continuity of outlines by splashes of white*'.

Kerr's letter to Churchill was not acknowledged until July 1915, but the Admiralty did act on his recommendations. He was told in December 1914 that his ideas were 'communicated to the fleet – confidentially in a general order'. He received confirmation of this from a former student of his, at that time serving in the Navy, who wrote that Kerr's scheme 'has aroused great interest on board among my fellow officers, all of whom I may say heartily endorse your statements, especially with regard to the increased difficulty of accurate range finding'.

But Kerr received little or no official feedback and became anxious that his ideas, left to individual ships' captains, were not being properly applied. In the summer of 1915, during the university vacation, he offered his services to the Admiralty to supervise the camouflage work, writing: 'I am doubtful from what I have seen of one or two vessels recently whether they are really making the most of the possibilities.' He wrote again to Churchill, suggesting a competition between ships for the best camouflage, to be judged – naturally – by himself. But in July 1915 the Admiralty informed him 'that various trials had been undertaken and that the range of conditions of light and surroundings rendered it necessary to modify considerably any theory based upon the analogy of animals'. A decision had been taken concerning the best scheme for colouring (a monotone grey), and that was the end of the matter.

Yet Kerr kept up his barrage of letters, sounding just as dogmatic as Thayer: 'It will be realized that what I have written above are mere statements of scientific fact.'

In June 1916 Kerr wrote to Churchill again, flattering him for having been interested in Kerr's suggestions for obliterative coloration but lamenting 'an almost complete failure to profit by your initiative'. By this Kerr meant that his idea on ship camouflage had been circulated to captains, who were left free to interpret them as they pleased. He then turned his attention to aircraft, recommending to Lloyd George that the undersides of aircraft be painted black, and not just any old black but the 'finest black velvet' for night attacks and white for day. But by that time no one was listening to Kerr any more.

Thayer was even more agitated than Kerr by the coming of the war. With America still two years away from entering the conflict, he turned to the British War Office, urging the staff to adopt his camouflage methods on both land and sea. In February 1915 Thayer wrote to Churchill, following in Kerr's footsteps. He wrote excitedly about submarines being rendered as ghostly as a shoal of fish by painting them like a countershaded mackerel. He followed this with a plea to paint surface ships white.

As with Kerr, Thayer's ideas entered the labyrinth of Admiralty memos. They were not ignored. Thayer had a friend at the War Office, a Colonel Bernard James, and he sought the latter's help too. And there was another friend, the American painter John Singer Sargent (1856–1925), who was then living in London.

The strangest of Thayer's ideas was inspired, not by fish or animals of any kind or by ships, but by some photographs of Bermudan houses taken in the sun:

They present to you exactly the aspect of the pure white tent that I proposed to stretch over your merchantmen, from the rail to the tops of their stacks . . . The angle of these imitation roofs on the merchantmen could easily be copied from the average thatch-angle of the Bermuda roofs (which, I think, is about that of the English thatches).

The verdict on Thayer's and Kerr's proposals was given by Captain Thomas Crease, naval assistant to the Admiral of the Fleet, Lord Fisher:

Several of these freak methods of painting ships have been tried – and on similar scientific principles. But the fact is that the requirements differ every day depending on the light, the colour of the sky and sea, and especially the time of day and position of the sun . . . I think the proposals are of academic interest but not of practical advantage.

This was to be the official line for two years; but, as 1915 wore on, Thayer became distressed by the stonewalling he received from Britain. His letters to Colonel James and Sargent became more manic. Finally, in November, he decided to go to Britain to campaign personally.

So began a farcical game of hide and seek. Thayer left for England on 13 November and, once there, hared around the country giving demonstrations but never leaving an address. Sargent was unable to contact him. In the words of his biographer, Thayer had 'partly lost his grip'. In Glasgow he met John Graham Kerr and was thrilled to find a sympathetic hearing at last. The state of his mind can be judged from the letter he wrote to friends in Oxford: 'total heavenly triumph'. Thayer and Kerr were in some respects birds of a feather, and the meeting of like minds was so successful that it totally deflected Thayer from the real purpose of his visit: convincing the War Office. Sargent eventually made contact and urged him to come to London, where the War Office was ready to discuss his plans. But Thayer preferred to leave on the high note of his triumph in Scotland rather than endure a grilling from the War Office. Thayer always demanded total assent and could not cope with any kind of challenge. He sailed for home. Sargent wrote to him in January 1916: 'I shall be curious to know what made you vamoose.'

But in March Thayer was once again pressing the Admiralty for action on marine camouflage. The First Lord, Arthur Balfour, who had succeeded Churchill, replied on 23 March 1916:

> The First Lord thinks that possibly you have overlooked the fact that the firing of guns in a fleet action is not directed from the deck or from any position approximate to sea level, but from a position at least 100 feet above the deck and by an officer equipped with the most powerful glasses.

In August 1916, after further unsuccessful pleas to the War Office, Thayer wrote an article in the *New York Tribune* urging the authorities to think again. Grandiosely, he reports that one English professor had urged all the naturalists to petition the government to adopt Thayer's methods, saying to Thayer 'your book has convinced us all'. Thayer breathlessly makes the case again for *white white white*.

An artist now enters the story who, it seems, was unaware of camouflage in the natural world. This was Norman Wilkinson (1878–1971), a marine painter and illustrator who, before the war, was working mainly for the *Illustrated London News*. Wilkinson was an old-fashioned painter who loved boats: come the war, he decided to join the Navy – which he did, performing the role of Paymaster from June 1915. He served on various fronts, including the Dardanelles, Gallipoli and Gibraltar, on submarine patrol. He returned to England in 1917 as a lieutenant in the Royal Naval Volunteer Reserve at Devonport, commanding an eighty-foot motorboat used for minesweeping duties.

In the spring of 1917, Wilkinson took a weekend's leave to go trout-fishing at Honiton, Devon. It must have been an inspiring break, because his many experiences at sea caused an idea to germinate in him. He returned to Devonport dockyard on an April Monday morning and went straight to his commander with the idea:

> . . . absolutely a spontaneous one, it simply arrived. I have been a sea painter as an artist and have studied the ships at sea all my life practically, and I was in this motor launch at Devonport, on patrol at that time. All the transports in the Admiralty service were painted entirely black and I

thought of this for a long time and I knew it was utterly impossible to render a ship invisible and it seemed to be something could be done on other lines, but it occurred to me quite spontaneously that the original idea was if you could break up this black surface with white in such a way that the course of the ship might be upset, not simply haphazard.

Breaking up a black surface with white suggests the zebra, and Wilkinson's designs, although they were tailored to particular boats, were almost all variations on black-and-white striped patterns, sometimes with blue or green, not curved as in the zebra but rigidly geometrical. The ability of black and white patterns to confuse the eye is even better understood now than it was in Wilkinson's day, thanks to the Op-Art of the 1960s and especially to the paintings of Bridget Riley. In a trial-and-error way, Wilkinson was making experiments in visual perception.

There is some doubt as to whether Wilkinson was aware of Thayer's and Kerr's ideas at the time. After the war, he wrote: 'I had heard of nothing at all beyond the old invisibility-idea which everybody knew, which, I believe, is in the first book of the Gallic Wars.'

The idea of invisibility, as opposed to unrecognisability or confusion of aim, was to become a pivotal matter after the war, when various claims for the invention of dazzle painting were assessed. If not invisibility, then what? Wilkinson's idea was that disruptive colour could make an assessment of a ship's course difficult for a submarine aiming to fire a torpedo after a sighting by periscope. It was not intended to help a ship to avoid long-distance gun attack, which had been Kerr's main aim. Indeed, Wilkinson admitted that dazzle painting might make a ship more visible in some circumstances, and hence more vulnerable. He described thus the rationale behind disruptive patterning and confusing a submarine's aim: 'The whole aim was to use perspective and colour strength in such a way that the bow was pushed away from you and the stern was brought towards you.' In other words, the patterns had to create an optical illusion almost like the impression of motion you feel in a stationary train when a train moves off beside yours.

If foiling submarine attack was the purpose, the ships to be protected, in Wilkinson's view, were to be merchant ships only. Although camouflage was later applied to warships, Wilkinson did not advocate it.

His idea was welcomed at Devonport and on 27 April 1917 Wilkinson submitted his plans to the Admiralty. Unlike Kerr, who spent his life knocking on doors that refused to open, Wilkinson had a lock-picker's knack. He had contacts both in the Navy and in the art world. The Admiralty authorised him to set up a camouflage unit but could not provide premises or staff. Once again pulling strings, he enlisted the help of the Royal Academy, which offered to provide a home for the new unit.

So Wilkinson set up shop in the Royal Academy in Piccadilly, London. He recruited some fellow artists who were all given Royal Naval Volunteer Reserve (RNVR) commissions, and the RA provided about twenty female students to help. Wilkinson needed to move quickly because in 1917 losses of merchant shipping to U-boats were becoming unsustainable, but the Admiralty was dragging its heels again, waiting for reports on the viability of Wilkinson's trial design. Thayer and Kerr's schemes were still on file: 'proposal by A. H. Thayer to paint submarines like high swimming open sea fish, a form of blue paint was tried'; 'proposal for diminishing visibility by compensating shades destroying the continuity of outline, etc – this scheme proved of little or no advantage (Kerr)'. After some more canny lobbying by Wilkinson, the Controller General of Merchant Shipping (CGMS) gave an order for fifty ships and took Wilkinson's unit officially under his wing, as the CGMS Dazzle Section.

Perhaps Wilkinson's strongest suit was for getting things done. And he was an insider, a practical seaman. To read Kerr's letters is to sense an over-anxious outsider, blinded by his own big idea, not quite knowing whom he was addressing. Whatever the reason, Wilkinson made dazzle the dominant mode, superseding the ideas put forward by Kerr.

W. F. Welch, who worked as a liaison officer between the Admiralty and Wilkinson's teams, described Wilkinson's technique:

> He had a very skilled modeller, who used to carve out models of ships and these models were placed on a revolving table and viewed through a periscope at the Royal Academy with suitable sidescreens and back grounds and lights were adjusted to get the various day and night effects.

Artists recruited included the vorticist painter Edward Wadsworth, working in Bristol and Liverpool. Wadsworth's role is unclear, but the relationship between the dazzle work and his own paintings is striking. It is paradoxical that Wilkinson was an old-fashioned realist painter, but his dazzle designs struck a chord in the vorticist artists, who were applying bold, energetic, stylised patterns to their subject matter. Wadsworth was not especially interested in some of the more complex dazzle patterns: what he wanted was stripes. He is best known in the dazzle context for a painting he made in early 1919: *Dazzle Ship in Dry Dock, Liverpool*. This was the first painting he had produced since the war. During the war he had contented himself with woodcuts, some of which related to his dazzle work.

By October 1917 Wilkinson's operation was sufficiently well established for King George V to pay a visit. The King prided himself on his maritime expertise and tried to guess the course of one of Wilkinson's model ships. He was incredulous when he was told that his estimate that the ship was travelling south by west was wildly wrong: it was actually east south east, proving the success of the exercise.

When America entered the war on 6 April 1917 (partly as a response to the sinking of neutral US ships by U-boats), there were already six approved ships' camouflage schemes in force. The rationale for the patterns was mixture of reduced visibility and disruption. One pattern – Brush – stemmed from Gerome Brush, Thayer's friend, and this indeed exhibited Thayer's principles of using countershading for concealment, but there is no evidence of the technique being used on a ship.

All this changed when, early in 1918, Norman Wilkinson visited the USA to advise. Here he met the Assistant Secretary to the Navy – Franklin Roosevelt, a fifth cousin of Theodore and later to be the great president of the Depression era and of the Second World War. Wilkinson was not impressed by the US efforts up to that point: 'When I got over there I found they had five men farming invisibility schemes at royalties of about 100 dollars per foot run, that was all on invisibility.' The emphasis on 'invisibility' stemmed from Thayer, who was convinced that, at night or under cloud, a white superstructure could make a ship invisible. In his rambling, repetitive, 1916 *New York Tribune* article he had claimed that, 'In cloudy weather (which secures uniform illumination) vertical pure white so counterfeits the sky against which you see it that only in the brightest moments is there any hope of distinguishing it from the sky.'

Following America's entry into the war, Thayer had begun to pester Roosevelt with his schemes. On 7 July 1917 he wrote to Roosevelt complaining about a vertical zebra stripe camouflage he had seen in a newspaper report concerning a submarine. Thayer urges pure white for verticals, as ever, but admits: 'Since the *pure white* vertical parts are too bright when sunlit or moonlit, the submarine can be provided with coarse black nets to haul up over these vertical parts when their spectator sees them thus illuminated.' Roosevelt replied a week later saying that the nets are 'quite impracticable'. He also pointed out, as the British Admiralty had to Kerr, that any pattern would only work in one set of lighting conditions.

Further letters from Thayer received stonewalling responses from Roosevelt. In April 1918 Thayer tried a different ploy: to erect billowing covers around a ship. Roosevelt replied: 'It is noted that the basic idea of your method involves the disguising of a vessel so that she would have the appearance of a cloud.' After repeating that this was impractical, he informed Thayer that the US and British navies were using the dazzle system.

When Wilkinson met Roosevelt, the latter was no more impressed than Wilkinson by the invisibility schemes; but Wilkinson got his hearing. A very high-ranking US admiral was given a demonstration of dazzle by Wilkinson and invited to guess a model ship's course. Like King George V before him, he got it wrong. He blew up: 'How the hell do you expect me to estimate the course of a God-damn thing all painted up like that?' Which, of course, made Wilkinson's point very neatly, as everyone who saw it observed.

The American artist Everett Warner (1877–1963) recalls how, working with Wilkinson, he became convinced that the use of dazzle patterns to confuse torpedo range finding was the only scheme worth pursuing. Aiming a torpedo, which travelled at a mere 35 knots (40 mph; 72 kph), involved guessing where the target would be when the torpedo arrived. Many US ships were formerly German ships whose speed was known to the enemy, so disguise of identity, as well as confusion of aim (both of which could be achieved by the same dazzle pattern), were desirable.

Warner recalled how, initially by trial and error, the US team discovered the rationale for course disruption by dazzle (Fig. 6.1). The patterns

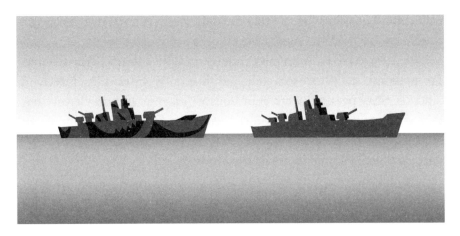

Fig. 6.1   Dazzle painting of ships in the First World War. The two ships are on the same course, coming closer to the observer, but the bow of the dazzled ship seems to be turning away, out to sea. (Adapted from Everett Warner, *Transactions of the Illuminating Engineering Society*, 21 July 1919.)

created the illusion that the ship was split into blocks semi-independent of each other. In this way, the ship could be made to look as if the bow were turning away even when it was pursuing a straight course. Wilkinson believed that this is a trick which nature does not possess – a point he would make later, when the priority for the invention of dazzle was disputed after the war. But he was forgetting the false heads and eyespots on some creatures, which are intended to deflect attacks.

Wilkinson was allowed to run the US ship's camouflage programme until London became restless and called him home after about five weeks. On his return, Wilkinson encountered his first big setback. Merchant captains almost all approved of the scheme and felt that it improved morale on the ships. But some captains in the Mediterranean reported that the use of white paint was dangerous, making the ships unusually vulnerable on moonlit nights. On 21 July 1918 the British Commander in Chief, Mediterranean Station, wrote to the Admiralty asking for dazzle to be withdrawn in his sector. The daylight conditions in the Mediterranean, the fact that most of the ships travelled in convoy (where the synchronised movement of the body of shipping was easy to detect) and their vulnerability at night, when 70 per cent of attacks now occurred, were all factors. Wilkinson was unrepentant: 'while it might have been somewhat visible it did not alter the fact that the difficulty remained of estimating a vessel's course from the submarine officer's point of view.'

The scare led to an Admiralty enquiry into the efficacy of the scheme. To preempt it, Wilkinson solicited reactions from fifty merchant captains. Sixty per cent were enormously in favour and only a very few were absolutely against it. An official report on dazzle-painted merchant ships was produced in September 1918. To the end of June 1918, 2,367 ships had been painted in the dazzle mode.

The statistics were inconclusive and could be made to fit whatever outcome one wanted to prove. For example, in the first quarter of 1918, a greater percentage of dazzled ships attacked was lost or damaged (72 per cent, against 62 per cent for non-dazzled). This was pretty damning, but then in the second quarter 60 per cent of attacks on dazzled ships resulted in sinking or damage as against 68 per cent of non-dazzled. And more dazzled than non-dazzled ships were attacked – 1.47 per cent of sailings in the six months against 1.12 per cent for non-dazzled. But Wilkinson could argue, as he did, that his scheme was not designed to make a ship hard to see but hard to hit. Consistently, fewer dazzled ships seemed to be sunk rather than damaged: 43 per cent over six months against 54 per cent, and 41 per cent of dazzled ships were hit amidships, against 52 per cent non-dazzled, which might be taken to confirm the difficulty in aiming at a dazzled ship.

But not much could be made of these figures. The comparison was made harder by the fact that the dazzled ships were much larger than average: 38 per cent were over 5,000 tons, whereas only around 13 per cent of non-dazzled vessels were over 5,000 tons. Captured U-boat men mentioned no difficulty in targeting dazzle ships. The report concluded:

> No definite case on material grounds can be made out for any benefit in this respect from this form of camouflage . . . At the same time the statistics did not prove that it is disadvantageous, and in view of the undoubted increase in confidence and morale of officers and crews of the Mercantile Marine resulting from this painting, . . . it may be found advisable to continue the system.

The problem of white paint in the Mediterranean theatre was noted and No. 1 grey substituted. Everyone, including Wilkinson, recognised that superstructure was the problem. Where a ship had substantial super-structure at either end, Wilkinson's attempt to make the bow recede and the stern approach was thwarted. Before much could be done to rectify these faults, the war ended.

The results of ship camouflage in the USA seemed more conclusive than in Britain. A total of 1,256 ships, merchant and fighting, were camouflaged between 1 March and 11 November 1918. Ninety-six ships over 2,500 tons were sunk; of these, only 18 were camouflaged and all of them were merchant ships. None of the camouflaged fighting ships was sunk.

After the war, an Admiralty enquiry was mounted into the question of priority for the invention of dazzle painting. There were several claimants besides Wilkinson, the most serious rival being John Graham Kerr. The committee began to meet on 27 October 1919. Wilkinson's testimony is straightforward: he had only one end in view, the disruption of aim of submarine range finders intent on firing torpedoes.

Kerr was a different matter. He had been cold-shouldered in 1915, only to see something which, he thought, looked like his idea suddenly re-appear in 1917. He was particularly troubled by the talk Wilkinson gave after the war at the North East Coast Institution of Ship Builders on 10 July 1919. This talk had been printed as a pamphlet and was documentary evidence at the hearing. In his submission to the committee, on 27 November 1919, Kerr had assembled quotations from his own work and from Wilkinson's talk which did seem to show a very close resemblance between their ideas. The naval panel, however, put it to Kerr that, while Wilkinson had aimed to disrupt the aim of submarine range finders, Kerr's proposal aimed at invisibility. Kerr admitted that he had not explicitly stated that the aim was to disrupt range finding, but he argued that such an aim could be inferred from his writings.

What is fascinating about the case of Wilkinson and Kerr is that, while they were concerned with the questions of visual perception involving *images* of ships and animals, they become embroiled in a war of *words*, in which verbal rather than visual distinctions were stressed. Much time was spent on the semantic relationship between 'invisibility' and 'unrecognis-ability'. Kerr's idea was to break up the outline and to destroy identity: 'of course it has a relation, but not a very close relation to invisibility'.

It is hard not to feel sorry for Kerr. There are indeed very close textual similarities between Kerr's letters of 1915 and Wilkinson's talk of 1919. For example: on 25 June 1915, Kerr had written to Churchill of 'stultifying the German range-finders'; on 28 June 1915 he had written to Balfour recommending 'obliterating those sharp details which are made use of in range finding'. This is very close to Wilkinson's principle.

And in his 1919 talk Wilkinson had seemed to endorse the Thayer/Kerr advocacy of white paint: 'For a time in the early stages of the scheme I hesitated to use white paint for various reasons, but after considerable experience, it was found to be the best "colour" for those parts of the ship intended to be invisible.' The man who didn't believe in invisibility had uttered the words 'intended to be invisible'.

As for Kerr, perhaps his fatal flaw was to think that his scheme could be useful against long-range guns: 'I was thinking entirely of the long range gun firing, I was not thinking of Submarines in those days' – but Wilkinson only spoke of submarines. Kerr obviously retained a hankering after invisibility, saying plaintively: 'that was called the Thayer system, that has not been given a fair trial, . . . it is merely a question of brilliant white paint'.

Given that Kerr had made his claim to the committee on the lines that he should be given credit for informing the Admiralty in 1914 of the scheme that was eventually adopted in 1917, he was asked: 'Do you suggest that he [Wilkinson] personally derived any advantage from communications made by you?' Kerr answered in a roundabout way, implying no. He didn't claim any personal credit for his ideas: 'the whole of this idea is a purely biological principle'. Any biologist, he claimed, would recognise the importance of countershading. 'I make no claim to have invented the principle of parti-colouring, this principle was, of course, invented by nature . . . even savages have used parti-coloured shields.'

It was put to him that 'no where in your letters do you suggest that your system of part-colouring would tend to create an illusion as to the course of the vessel painted'. Kerr had to admit that this was true. The Lords Commissioners of the Admiralty ruled: 'Incidental resemblance is no ground on which a claim can be properly based when the specific purpose referred to was not actually put forward by the claimant in his original proposals.'

Kerr was informed by the Admiralty in October 1920 that he was 'not regarded as being responsible for the adoption of that system of painting' [dazzle]. He was advised, if he wished to pursue it further, to apply to the Royal Commission on Awards to Inventors. So began another enquiry, with legal representation.

The records of the Royal Commission show that its members were not very impressed by either claimant, saying of Wilkinson:

> The effect of 'Dazzle Painting' as a protection against submarine attack proved to be disappointing in practice. As a result of all available statistics it appears that the degree of immunity conferred against attack or injury as between dazzled and non-dazzled ships is very slight.

As regards Kerr, they recognised his priority but felt that his proposal was 'for variegated painting of ships to procure invisibility', while 'the method actually adopted was directed to an entirely different purpose'.

During the hearings, Kerr had been engaging with Wilkinson in the pages of the science journal *Nature* and *The Times*. In his letters, Kerr tries to claim priority: 'newspaper paragraphs which date the discovery of the principle ... from the year 1917 are distinctly misleading'. He could not bring himself to mention the name of Wilkinson at all.

Kerr's letter received a devastating riposte from the unacknowledged Norman Wilkinson just over a month later. While Kerr believed that nature could teach everything a naval *camoufleur* would need, Wilkinson retorted that his goal was to confuse an aimer with respect to the course of a ship; he tartly observed: 'I am not aware that this occurs in biology, ie disguise of direction.' Nobody reminded him of butterfly eyespots.

In 1922 Wilkinson was awarded £2,000 for his invention. Professor Kerr continued his zoological work and became the mentor of Hugh Cott – both of them would be involved when questions of camouflage were once again high on the agenda. Kerr never accepted the verdict of the enquiry and the Royal Commission and would spend the rest of his life seeking vindication.

Wilkinson, Thayer and Kerr were very contrasting figures. Wilkinson was calm, practical, bluffly unintellectual and, above all, lucky. Thayer was tormented, neurotic, unworldly – he had all those traits that bad luck always battens on. Kerr was stubborn and blinkered – an inflexible man with an *idée fixe*. Wilkinson was affable and confident, with an easy charm. If he had an idea or a problem, there was always a well-connected friend to open doors for him. He was at ease with kings and future presidents of the USA. He enjoyed what of the high life came his way in a simple, uncomplicated manner. In his memoirs, he often didn't remember the names of the people he had met. He was a traditional marine painter – an old-fashioned, four-square, full-masted, honest-jibbed fellow. Yet there must have been something in the air in those days – not just of war, but of modernist ferment. Because this most traditional of painters produced designs for ships which were a glorious success as avant-garde art. Given their scale, they must be counted as some of the most dramatically successful artworks of the period.

To look at the ships now is to see Op-Art in action, applied not to the mini-dresses, coffee tables and curtains of the 1960s but to ocean-going hulks of metal. The designs chimed perfectly with the vorticist movement. This potential was realised as soon as the war was over in the Dazzle Ball, held at the Chelsea Arts Club on 12 March 1919. The *Illustrated London News* carried a jazzy and exuberant piece of artwork under the heading: 'THE GREAT DAZZLE BALL AT THE ALBERT HALL: THE SHOWER OF BOMB BALLOONS AND SOME TYPICAL COSTUMES.' The caption read:

> The scheme of decoration for the great fancy dress ball given by the Chelsea Arts Club at the Albert Hall, the other day, was based on the principles of 'Dazzle', the method of 'camouflage' used during the war in the painting of ships to help them in escaping from the attacks of submarines. Many of the costumes were also designed specifically for the occasion on 'Dazzle' lines, but there was also a great variety of fancy dress of the ordinary type. The total effect was brilliant and fantastic. During the evening a shower of 'bombs' in the shape of coloured balloons descended on the devoted heads of the dancers, and added greatly to the hilarity of the occasion.

This adoption of military camouflage for adornment prefigured the eruption of 'camo' fashion from the 1960s onwards.

CHAPTER 7

CAMOUFLAGE AND CUBISM IN THE FIRST WORLD WAR

In order to deform totally the aspect of the object, I had to employ the means that cubists use to represent it.

Lucien-Victor Guirand de Scévola, artist and
First World War *camoufleur*

If zoologists had at least put ideas forward in the war at sea, land camouflage was almost completely dominated by artists. Indeed, the distinguished marine biologist Sir Alister Hardy commented: 'we should note that the successful camouflage officers of the two world wars were artists rather than scientists, or sometimes scientists with artistic inclinations'. His own experience seemed to bear this out: 'I was chosen as a camouflage officer in the first world war by Solomon J. Solomon, R.A., under whom I received my training, through, as I learnt afterwards, his mistaking me for another Hardy, a professional artist who had been a pupil at Herkomer's school!'

Ground combat is the closest human counterpart to the struggles in the natural world which gave rise to all the forms of deception in which we are interested. Troops roam about the same natural environment as the animals: they seek and evade, ambush, create decoys. The big innovations of the First World War over earlier wars were the machine gun, the aeroplane and the tank, and these innovations had consequences for camouflage. The aeroplane was the most important. Right from the start, aeroplanes were used for reconnaissance as well as for fighting. The need to hide large concentrations of troops and equipment from the spy in the sky became urgent. And the machine gun pinned down large formations in trench warfare, leading to such fascinating deceptions as the dummy tree look-out posts, and rows of wire-manipulated puppet 'Chinese' troops.

Camouflage in the First World War essentially meant concealing equipment rather than formations of troops. Of course, the idea of making soldiers invisible was an attractive one, but disguising a single sniper was much more likely to be productive than decking out a whole army in camouflage. There are several reasons. Snipers rely absolutely on concealment, whereas regular troops depend more on firepower. The sniper's art derives from hunting, in which camouflage has often been used in stalking animals. And mass producing camouflage uniforms was not easy technically. Sniper suits could be, and were, handmade.

But such arguments were lost on Abbott Thayer, who, when war broke out, felt that he had a mission to help the Allied cause on land as well as at sea. His idea of painting ships white was daringly counter-intuitive, but his proposals for camouflaging troops on land were no different from those which had been used by deerstalkers for generations. A camouflaged uniform must blend in with the natural textures of the environment. Patterns of disruptive colour to break up the characteristic human outline can help, but scraps of loose fabric and foliage are better for creating something that looks more like a shrub than a man. This is ideal for a lone sniper but might not be appropriate for front-line fighting troops.

Thayer's ideas for army uniforms did indeed include scraps of loose cloth. He had sent some samples ahead of him on his urgent visit to Britain in November 1915. The day before Thayer had left, his friend John Singer Sargent had written to him a letter which Thayer never received until he returned home; the letter was telling him that the War Office had considered his uniform camouflage, but, 'although they admitted that your observations of nature and effects were "gospel truth" as they said, there is little likelihood of their ordering in large quantities anything resembling your uniforms'.

Thayer was informed of this personally much later, in August 1916, when the Director of Equipment and Ordnance Stores wrote to him:

I had many suits made up to your pattern, and also of Khaki of a mottled design similar to that of which you showed me sketches. Whilst appreciating the accuracy of your contentions, it is found that to apply it to the use of soldiers en masse, whether for scouts or otherwise, is not practicable.

This chapter of Thayer's campaign came to a close on 3 October 1916, when he wrote to the War Office to terminate the correspondence, having given up in disgust. In Thayer's archive there is a tantalising last entry in his War Department files. It is undated and apparently from his friend Viscount Bryce (1838–1922), who had been British Ambassador to the USA from 1907 to 1913:

I am sending you this to show that your teaching *has* borne fruit, and that our British soldiers are protected by coats of motley hue and stripes of paint just as you suggested . . . The tanks also are painted in protective coloration just as you would wish.

This must be a reference to the sniper suits, since no camouflaged uniforms were issued to regular troops in the First World War. It is true that, from 1916 on, snipers were wearing handmade camouflaged smocks adorned with tattered strips, very much like Thayer's designs. In fact, these did not derive directly from Thayer but were based on the traditional costume of the Scottish deerstalkers known as 'ghillies'. The latter were not much different from Thayer's samples but the inspiration clearly came from the Highlands of Scotland, not from New England.

It is understandable that some naturalists instinctively felt that they, and they alone, could lead the military towards the successful application of camouflage. But the First World War erupted in the middle of the biggest sea change in art since the Renaissance. In the first decade of the twentieth century, without any knowledge of disruptive coloration, artists were experimenting with breaking up the outline of figures and objects in their paintings. This was most extreme in cubism. The muddy, multifaceted diced guitars and human beings painted by Picasso and Braque around 1910 already look like camouflage, something that was not lost on the creators themselves. Gertrude Stein tells the story of how she was walking with Picasso on the Boulevard Raspail in Paris at the beginning of the war when a camouflaged truck passed. Picasso was amazed and exclaimed: 'Yes it is we who made it, that is cubism.'

'It is we who made it,' was typical Picasso hyperbole. But it was echoed by his co-cubist Georges Braque; in an interview in 1949 Braque recalled:

> I was very happy when, in 1914, I realized that the Army had used the principles of my cubist paintings for camouflage. 'Cubism and camou-flage,' I once said to someone. He answered that it was all coincidence. 'No, no,' I said, 'It is you who are wrong. Before cubism we had impressionism, and the Army used pale blue uniforms, horizon blue, atmospheric camouflage.'

Was there any truth in it? In the First World War the French Camouflage Corps was the first to form; it set the tone for the rest of the Allies. The leading light was the painter Lucien-Victor Guirand de Scévola (1871–1950). The French operatives were called *camoufleurs*.

Most of the artists involved in camouflage were traditional representa-tional painters. But, as de Scévola's epigraph to this chapter shows, some were aware of cubism. De Scévola was married to a *comédienne* at the Comédie Française, and he was 'familiar with the world of painters, cubist or not, and theatre designers specialising in *trompe l'oeil*'. Theatrical illusion was obviously a close cousin of military deception.

There had been a convergence between the disruptive patterns urged by Thayer and Kerr and the parallel disruption of the image practised by cubism. But between the zoologists and the painters there was no under-standing of this rapprochement. Thayer was violently opposed to avant-garde art; Wilkinson never expressed any opinions on the subject, but his own painting and his whole sensibility – club, boats and parlour songs – were Edwardian. There is no evidence that Picasso and Braque were influenced by the studies of naturalists. In the 1940s the surrealist painter Roland Penrose recognised that 'the point they [cubism and camou-flage] had in common was the disruption of their exterior form in a desire to change their too easily recognised identity'. Penrose was also well versed in nature, but in the First World War the links between nature, cubism and camouflage were imperfectly realised, however obvious they might appear in hindsight.

Perhaps because painting was in such ferment, the early experiments in camouflage are often garish and, to our eyes, more successful artistically than as concealment. The problem was the idea of disruptive coloration. Breaking up the outline of any form is not effective as camouflage if the object is still highly conspicuous. Later on in the war, Picasso remarked to Cocteau: 'If they want to make an army invisible at a distance they have only to dress their men as harlequins.' It is not clear whether he was being insouciantly provocative or serious, but in fact many camouflage schemes in the First World War were bright, gaudy, harlequinesque disruptive patterns rather than the muddy earth tones to which we have since become accustomed.

Whatever the theories and background, there was work to be done. De Scévola began in September 1914 by concealing a gun emplacement using painted canvas. In February 1915 he was given a small experimental setup at Amiens. The first notable piece of work was the construction of an observation tree erected in May 1915. Observation trees became a fascinating staple of First World War camouflage, the idea being to remove a blasted tree trunk at night and replace it by a steel lookout post disguised with bark to replicate the real tree.

De Scévola's work thrived and he was given command of the Camouflage Corps in the Autumn of 1915. André Mare was one of the artists employed under de Scévola, and he kept a sketchbook of water colours. Mare's initiation was rapid: from believing that the goal was to paint flowers on cannons and vehicles and to paint the horses white, he realised that the breakup of form and colour undertaken by the cubists from about 1908 couldn't have been more timely: camouflage, in breaking the outline of cannon and other war machinery, would take its cue from cubism. Although the French *camoufleurs* took as their emblem the chameleon, they were artists, not naturalists.

Mare specialised in creating tree lookout posts. He had some scary moments, cloistered within a tree observation post during battle, his feet in a metre of water. 'It's no cushy number, camouflage,' he wrote; 'everybody out there thinks it's a sleepy waiting game and when you tell them what it's really like it throws them a bit.' Mare's cubism is apparent in his wartime sketchbook.

At this point, the British became interested and the painter Solomon J. Solomon, R.A. (1860–1927) was sent to France in December 1915 to set up a small camouflage unit. The story of Britain's first army *camoufleur* has strange echoes of Thayer and Kerr. Solomon came from a South London Jewish family and was a traditional artist, sometimes described as a Pre-Raphaelite, painting portraits and mythological scenes, such as his best-known work *Samson* (1887). Alma Tadema and Frederick Leighton were influences. Like Thayer, he did not welcome cubism, vorticism, or any other form of modernism.

When war broke out in August 1914, Solomon began experimenting with camouflage screens made from dyed butter muslin and bamboo poles. A letter to *The Times* on 27 January 1915, advocating camouflage, suggested that he was aware of Thayer: 'The protection afforded animate creatures by nature's gift of colour assimilation to their environment might provide a lesson to those who equip an army.' This led in December 1915 to an invitation for Solomon to go to France to establish a camouflage unit. He was given a temporary commission as lieutenant colonel and was allowed to recruit his own small team.

In January 1916 the unit was installed temporarily at St Omer, while a derelict factory at Wimereux was prepared for them. The company consisted of Solomon, five artists and a civvie street carpenter. By March the unit was settling down at Wimereux. In case anyone got above themselves, 'the artists were rated for Engineer Pay as painters, and the sculptors as plasterers'. Solomon found himself referred to as 'Mr Artist'.

But the newly appointed commander in chief, Field Marshal Haig, was encouraging ('the whole of Flanders is at your disposal'), and de Scévola and the French *camoufleurs* – who had a good head start on Solomon – were helpful. Solomon got on well with other artists, but he never understood military protocol and the culture of 'everything goes through me'. The first sign of trouble to come is revealed in the detailed war-time diary he kept. The shed allocated at Wimereux was supposed to have been a minerals factory, but Solomon was certain it had been built by the Germans before the war as a post to be used during the campaign: this factory was 'nothing but an effective screen for a big gun' – presumably intended to fire across the Channel. To back up his claim, he cited stories about Germans living in England before the war who were rumoured to have built up arms dumps in caves, and he referred to the Wimereux site as 'the derelict factory put up by the Germans'.

Paranoid fantasies would be Solomon's undoing, but for a while things went well. De Scévola's French *camoufleurs* were making screens for gun emplacements from painted canvas, but there were suspicions that from the air this would not be convincing. Solomon quickly improvised netting with bundles of hay, tied into the mesh and coloured. Solomon wrote in his diary: 'It could be stretched over guns – dumps, trenches, and over quite large areas; it would make no appreciable shadow on the ground – in fact it was to be the scientific solution of most of our screening problems.'

This was the basis of what became a stock universal screen and, according to Solomon, 7 million square yards were used by the end of the war. The netting didn't come into its own until 1917; for the time being, what the army wanted was tree observation posts. These they learned to make from the French. A fuss was made over the first one, erected on 22 March 1916. The trees concerned were mostly willows and, for security reasons, Solomon decided that the best source of willow bark was the King's park at Windsor. So this first British observation tree featured 'real bark sewn on canvas and came from a willow in the King's park at Windsor'. That note of English patronage and snobbery sounds very odd in the context of the hellish landscape of mud and bodies which the post would command. Unfortunately, like many of the observation trees that followed, it was too constricted to be comfortable and too far from the front to reveal anything very special.

Solomon was a good *camoufleur*. He was undogmatic and let the logic of the situation dictate the camouflage concept rather than trying, as Thayer had, to harness an obsessive idea to inappropriate situations. For instance, the French tree observation posts were made round, but Solomon designed oval trees with the narrow end pointing towards the line, to give the impression of a tree too narrow to hold a man. Cricked necks must have ensued.

Despite the success of the trees and netting, Solomon's failure to understand the military setup led to him being replaced as head of the camouflage operation at the end of March 1916. Colonel Francis Wyatt of the Royal Engineers was appointed in his stead, and the whole operation expanded and put on a more professional basis. Solomon became technical advisor.

In May 1916, Solomon was asked to return to England to advise on camouflaging the new tanks that were being developed. He was sceptical about the possibilities of tank camouflage because of the shadow the vehicle would cast. Before he left, he met Haig again, and for some reason asked if he would be retained. Haig replied enigmatically: 'We must win this war.' On 1 August he was told he would be retained only in an honorary capacity.

Solomon spent some more time in France, but his tour of duty there ended in December 1916; he then founded a camouflage school in Kensington Gardens, on his own initiative. As with his first foray into camouflage, this too was eventually taken over by the army and regularised.

Under Wyatt, Solomon's netting became a staple item. As the importance of camouflage was recognised, attempts were made to improve the distribution to the front, but problems were never really solved. More successful were the ingenious 'Chinese' attacks. These were 'launched' by wire-operated dummy soldiers – plywood silhouettes – which danced on the skyline and attracted enemy fire, thus revealing enemy positions. At first the dummies were hand-painted on plywood, but later they were stencilled on millboard in ten standard patterns, representing 'the various attitudes that might be assumed by a man between the lying and upright position'.

The dummy soldiers were first used on 28 July 1917. Five machine-gun positions were revealed by the ploy and the dummies were left standing after the attack, to show the enemy he had been fooled and should 'look twice before calling on his artillery to put down a barrage'. Days later, a real surprise attack could be mounted from the same position. The official record of the Royal Engineers recounts with glee that, on 26 September 1917, 280 figures were 'all smashed and rendered useless by machine-gun, rifle and shell fire'. Such attacks, using around 300 figures, were common in 1917. In one case, a captured German report revealed that the Germans regarded it as a real attack.

Back in England, during 1917 and 1918, Solomon developed the military equivalent of Thayer's obsessive need to find camouflage in every living animal. An obsession with deception can sometimes be self-deceiving. Poring over some air reconnaissance photographs taken by his nephew, Solomon became convinced that the Germans had acquired the ability to conceal whole armies beneath vast nets propped up by poles high enough to allow the military traffic to pass beneath. On top would be painted the terrain observers would expect to see. He pestered the authorities but was ignored – until the great German offensive of March 1918, that is. This threw the Allies back to where they had been two years before, seeming to render meaningless the hundreds of thousands of lives lost in gaining that territory. Perhaps the Germans had sprung some surprises using camouflage? Solomon was asked to return to France in July 1918, although on a purely honorary and unpaid basis, to investigate his own claims.

Nothing came of Solomon's obsessive quest. The war ended in November 1918 with the Army never having accepted the idea of the vast German camouflage deception. Solomon returned to France and Flanders in 1919, to try to find evidence on the ground, but he believed that the Germans had covered their tracks. Nevertheless, he wrote a book, *Strategic Camouflage* (1920), detailing his mammoth German camouflage deception discovery. This was savagely dismissed in Britain, but some reviews in German publications gave him encouragement. A review in *Das Technische Blatt*, a supplement to the *Frankfurter Zeitung*, claimed that the large-scale camouflage was genuine: 'they were the very first things we destroyed on our retreat'. But then, if someone flatters you as the possessor of enormous powers, you might be inclined to agree with them, whether the powers were real or not. *Keep them guessing* was never a bad idea. The memoirs of the unofficial German Commander-in-Chief Luddendorff added further fuel by boasting of having concealed 40–50 divisions from the Allies. The truth has never been established.

It is sometimes claimed that Abbott Thayer did manage to influence the American camouflage effort through his cousin and artistic disciple, the mural painter Barry Faulkner (1881–1966). There is something in this, because Faulkner grew up worshipping Thayer as an artist and had absorbed his ideas on camouflage. He tells the story of Thayer coming to his school to talk about his ideas. Thayer brought his countershaded duck and was demonstrating it, when 'suddenly a frightened cat bounded between Thayer's legs, avoided the ungraded duck and dashed into and knocked over the duck it couldn't see. Cousin Abbott was as happy as a child at the cat's vindication of his theory.'

In 1917 when America entered the war, Faulkner and his artist friend Sherry Fry decided to become *camoufleurs*. No organised camouflage corps existed, so some advocacy was necessary. Faulkner went to Washington to lobby for a camouflage corps. Faulkner had no clear idea of how to apply Thayer's theory to camouflage, but he felt empowered by this arcane knowledge, believing that his association with Thayer gave him a 'bogus prestige'.

As it happened, the army had already realised the need for a camouflage corps and the Officers' Training Corps at Plattsburgh, New York State, was already on the case. A camouflage corps, officially called the 40th Engineers, was established. Surprisingly, Faulkner enlisted as a private. His company left for France on New Year's Day 1918. As he landed at Brest in Brittany, the ships in harbour were 'a riot of disruptive patterns', which must have been encouraging to someone just about to put Thayer's principles to the test. But at this late stage of the war, Faulkner found the reality to be confused and the possibilities for camouflage constrained:

> Our first attempts at camouflage were crude, and we had no air photographs to show us our mistakes. We surrounded the batteries of 75's and the larger guns with posts connected by heavy wires over which we stretched the camouflage material very much as the Connecticut Valley farmers cover their tobacco fields. The camouflage materials, made in our Dijon factory, consisted of rolls of chicken wire with multicoloured rags attached to them. The hard-boiled veterans of the First Division who at first jeered at camouflage and at us, the 'window-dressers,' grew to like it, for it gave them a sense of protection, false yet comforting, and in summer the rags shed a thin, pleasant shade like an unprosperous grape arbor.

The brutal realities of war often swept away the *camoufleurs'* efforts. On one occasion, going to inspect the positions he had camouflaged earlier, he found them suffering from heavy German attack:

At the guns we found . . . my camouflage in tatters, a gun pulled out of position and then abandoned . . . I hurried Homer [Homer Saint-Gaudens, Faulkner's commanding officer] to another camouflaged position well hidden in a grove of trees. On arrival, I barely recognized the place, for the position was a mass of blasted trees, its debris full of dead horses and dead men; it smelled foul. We beat it away from the gassy wood.

Despite the horrors, Faulker retained a characteristic New England innocence and, even in the landscape of blasted trees, mud and bodies, he managed to find some bucolic experiences. The trail of war led him back to Brest where, in its last days, he and a few friends boarded with a friendly French family with two gorgeous daughters. A bacchanal ensued, with good food and wine and 'plenty of frolicking in and out of the beds'.

Faulkner left France on Christmas Day 1918. He found Cousin Abbott in poor health, his nervous condition having worsened. Faulkner and his New England painter friends resumed artistic life as if the Armory Show and the cubism that had so disgusted Thayer had never happened.

Thayer's war has a postscript. In 1918, a new edition of *Concealing Coloration* was published with a fresh preface by his son, Gerald Thayer:

Is this a 'war-book'?
Yes and no.

The preface goes on to recount the story of Thayer's visit to Britain in 1915 and to claim the important influence his ideas had on the wartime camouflage effort. In a curious circularity of argument that we shall encounter again, Gerald Thayer both claims the influence of natural protective coloration on wartime camouflage and uses the stark test of war as a validation of his ideas concerning natural camouflage: 'War is no dilettante or trifling tester.' This was putting a brave gloss on Thayer's war work. In 1910, he had lamented:

My strenuous and essential life is art, and ornithology I only *border on*; and I have been so repeatedly disappointed by bodies of ornithologists seeming to me mainly unenthusiastic, unpoetical men, elbowing for self-display instead of merging self in worship of nature, that I have steadily drawn away from them, preferring to go into the woods, as it were . . . This way of feeling often shows one's own egotism, and very likely does in my own case.

Thayer's battles with the military authorities were even more bruising to him, and it seemed to him in his last years that he had failed. Abbot Thayer died in May 1921.

With the war over and the dazzle principle finding a purely frivolous use in fashion, the realms of zoology and art resumed their separate paths. The question of military camouflage remained vexed. The official account of the work of the Royal Engineers, written in 1926, made the admission:

The word *Camouflage*, associated as it is with dead horses, or pantomime, and savouring therefore of mystery and special technique, led, often unconsciously, to the adoption of an apathetic or non-serious attitude towards *camoufleurs*. However, the word has arrived with every appearance of making a long stay.

CHAPTER 8

HOPEFUL MONSTERS?

> There are, on your small wings,
>   black spots and splashes –
> like eyes, birds, girls, eyelashes.
>   But of what things
>   are you the airy norm?
>   What bits of faces,
> what broken times and places
>   shine through your form?

Joseph Brodsky, 'The Butterfly'

The years between the two world wars bred many monsters of one kind or another: Fascism, Nazism, Stalinism. It was a period in which everything seemed grotesquely distorted, a period when abstraction in visual art ceded to surrealism, when the scientific revolutions of relativity and quantum physics, and the technical advances of the motor car, the aeroplane and the radio were together producing dramatically unsettling effects on society. And the starting point for the 'twenties was the collapse of empires, economic chaos and the political bitterness which resulted from the First World War.

The war had ushered in the machine age of mass production and, in biology, the leisurely Victorian and Edwardian tradition of amateur field observation, which had played such a role in the story of mimicry, ceded to a new interest in the processes of life. There was less interest in collecting further examples of startling likenesses and more in trying to understand pattern formation in general, of which mimicry and camouflage were specific examples. One idea of pattern formation became known as the 'hopeful monster' theory.

The rediscovery of Mendel's work had opened a window into the mechanism of inheritance. And then there were Morgan's fruit flies and their bizarre mutations. In the 1920s the great revelation concerning the evolution of pattern-making in nature came in the field of butterfly wings. It was known as the 'Nymphalid groundplan'.*

How does nature create patterns on a butterfly's wings which can copy a leaf or the pattern of a different butterfly? Earlier naturalists were content to describe the acts of mimicry and to explain their existence in terms of natural selection gradually turning one pattern into another. But Darwinian natural selection is what is known as a 'black box'. We see what goes into the box and what comes out, but have no idea how the box transforms the input into the output. For most of us, our computers, TVs, mobiles and iPods are black boxes. We can manipulate them to obtain the results we want, but have no idea how these results are processed in the equipment. This is fine for consumers. But some biologists – from Bateson and Punnett onwards – have always asked the question: what happens within the Black Box of Evolution: what are the mechanisms that create these patterns?

If a butterfly-wing masterplan existed – one from which all actual patterns could be derived; a method in the madness of countless arrays of bands, whorls, friezes, rays and eyespots – that must surely be a clue to the processes behind the patterns? It would suggest experiments. In an example of simultaneous scientific discovery reminiscent of Darwin and Wallace both alighting on the theory of natural selection, a Russian and a German researcher discovered the butterfly pattern of patterns, the Nymphalid groundplan, independently, in the 1920s.

The two masterplan researchers were Boris Schwanwitsch (1889–1957) at the University of Petrograd in Russia and Fritz Suffert (1891–1945) at

---

* 'Nymphalid' refers to a large family of butterflies, the Nymphalidae, whose wing patterns are generally derived from the groundplan.

the University of Freiburg in Germany. Schwanwitsch was the first to publish, in 1924, but Suffert's very similar scheme is usually preferred. It seems amazing that anyone in 1924 in Petrograd (St Petersburg, as it was then and is again now) could focus on butterflies: Schwanwitsch did admit that his work had been delayed 'owing to the war and revolution'.

What was this groundplan? Butterfly wings have veins running from the root to the tip, which divide the wing into compartments or cells (Fig. 8.1). Running roughly at right angles to these veins are waves of

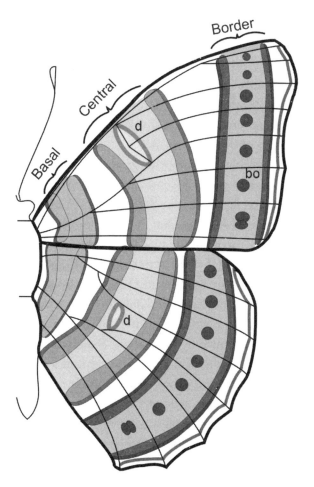

Fig. 8.1 The Nymphalid groundplan is the masterplan for many butterfly wing patterns. Most butterflies have some of these elements but not all and the features can be enlarged, reduced or distorted. (Illustration courtesy H. Frederick Nijhout)

patterning. Close to the root, these bands tend to be solid and to spread across the cells in curved lines. Towards the edges, each cell potentially contains an eyespot, fringed by a double border on both sides. Further fine borders skirt the wingtips.

Very few butterflies have the complete groundplan, and some such as *Heliconius* seem to bypass the pattern altogether; but very many have a partial groundplan, in which some elements are missing or distorted. The full plan can be deduced from the common elements shared by so many butterflies. There is no doubt that the plan is real and not merely a human projection of order onto chaos.

The groundplan immediately suggested ideas about butterfly wing patterns and how one species can apparently copy the patterns of another. It also immediately raised new questions. Punnett had suggested that there might be a limited range of patterns available to mimetic butterflies. Schwanwitsch and Suffert's work went one stage further: in effect *there might be only one pattern*, most actual ones being derived by omitting or distorting elements of the masterplan.

It was soon possible to apply the groundplan to one of the most striking examples of special resemblance: the *Kallima* butterfly. At first sight, nothing could be less like the groundplan than the leaf pattern on the underwings of *Kallima*. But in 1927 Suffert showed that, by suppressing some elements of the plan, highlighting others and holding the fore and hindwings so that the patterns run continuously across both of them, the leaf imitation is achieved (Fig. 8.2). Eyespots are (and are meant to be) conspicuous, so in *Kallima* they are reduced in size and function as mere 'fungal' blemishes – tiny 'holes' in the wing.

Perhaps the most astonishing thing about *Kallima*'s camouflage is that no two *Kallimas* have identical underwing colouring. Because dead leaves vary so much, *Kallima* wings range from 'bright yellow, reddish, ochre, brown, and ashy, just as leaves vary in their different stages of drying and decay'. The fake 'holes' and fungal blotches decorate the wings in appropriately random locations. Natural selection has not standardised the leaf pattern precisely because no two leaves are exactly alike. Being too precise a mimic would not work. Predators would begin to recognise the 'too-perfect leaf' as no leaf at all, and begin to seek it out. Although the ability to mimic so many kinds of leaves is striking, with

Fig. 8.2 How the *Kallima* butterfly mimics a leaf. The dominant aspect of a leaf is the midrib and *Kallima* achieves this appearance by a composite. On the forewing, the rib starts out as one of the main bands of the central symmetry system but this then takes a turn to the left and becomes a side vein. The continuation of the midrib is achieved by the border line of the reduced eyespots. The distortions from the groundplan to achieve this effect are striking and so is the suppression of more elements of the groundplan to yield an overall soft-toned brown leaf effect. (From H. Frederick Nijhout, *The Development and Evolution of Butterfly Wing Patterns*, Smithsonian Institution Press, 1991.)

hindsight and some genetic knowledge it is not so surprising. If only one leaf form and colour were in existence, natural selection would have eliminated those *Kallimas* which failed to match it. But, because leaves vary so much, 'loose' matching has been enough. This, of course, is a clue to the kind of genetic mechanisms involved: they must be correspondingly sloppy. Suffert had no idea of the deep genetic processes which had caused the groundplan to mutate into the leaf pattern mimicked by *Kallima*, but the same elements are so recognisable in each – there could be no doubt about it.

If the Nymphalid groundplan brought some order into the analysis of butterfly wing patterns, where should one start in looking for the mechanism? In the 1920s nothing was known about the chemical nature of the prime biological processes. Proteins were known as a category of substance, but their chemical composition was too complex to be deduced. DNA was known as a substance too, but its central role in inheritance, tissue-making and cell maintenance was completely unsuspected and, again, its chemical composition and structure were blanks.

If you have a black box whose inner workings are mysterious, how can you begin to investigate? You simply change the input signal and see what comes out. Obviously, your ignorance of what lies inside can lead to misinterpretation, but it's a start. This is the perturbation approach.

In the case of butterfly wing patterns, the simplest form of perturbation is what is called heat-shock or cold-shock. The pupae are treated with heat or cold for varying periods and the results produced in the adult butterfly are noted. The biologist to whom the cold-shock experiments spoke most strongly was the German-Jewish zoologist Richard Goldschmidt (1878–1958).

Goldschmidt was a colourful figure from a culture that now seems utterly remote. He wrote an autobiography in which his scientific work was relegated to a mere appendix. Goldschmidt loved music and art, and was a curious traveller in many countries, especially Japan and India. He was a professor in the old sense: a man of authority and self-assurance – so much so that his students called him 'the Pope'. He travelled in the grand style and his autobiography is littered with statements which now raise a smile, such as: 'A man over twenty usually does not know anything of fashions and leaves it to his tailor to clothe him properly.'

Goldschmidt's career was deeply skewed by the twentieth-century wars. He was in America when the First World War broke out and was interned as an enemy alien. He was forced to leave Germany for America again in 1935, with the rise of Nazi persecution. His rather distinguished family was personally maligned in a poster demonstrating its family tree. Goldschmidt countered the intended denigration with the remark: 'It could well be used as a chart demonstrating the effect of long selection of favourable hereditary traits upon the improvement of human faculties.' In America he became a professor at the University of California, Berkeley.

Goldschmidt echoed Bates in believing that butterfly wing patterns could be a model system for understanding pattern formation, because of their simple, flat nature, as opposed to the three-dimensional structure of most biological patterns, and because they are amenable both to genetical and to embryological experimentation.

For Goldschmidt, cold-shock or heat-shock experiments were immensely revealing about the link between the internal biology of butterflies and the environment. A large change in pattern could be produced by a simple environmental change. What more could the biological detective need in order to make some deductions about pattern formation and to devise new experiments?

Heat-shock and cold-shock applied to pupae are especially interesting because such procedures can sometimes duplicate the seasonal forms of the adult butterflies. A butterfly born in a green summer will often have a different appearance from that of the same species born in the autumn, when the leaves are turning. In general the summer forms are brightly coloured, with large eyespots. Their defence against predators involves the startle response of the eyespots.

Eyespots as a form of defence are an intriguing variation on the deceptions we have already seen. Like the snake's-head pose of the elephant hawkmoth caterpillar, they are a warning sign, a threat. But, unlike that sign and the ones used by many other toxic insects which advertise their toxicity, eyespots are the bluff of harmless creatures. No specific creature is being mimicked: it appears that large eyes have an alarming effect on many birds, so that large eyespots, suddenly flashed, can startle a bird into allowing an insect time to escape.*

A peacock butterfly, for example, has two main antipredator strategies. When its wings are closed it is an excellent leaf mimic, very much like the Indian leaf butterfly *Kallima*, to which it is related. But if this strategy fails and a bird is closing in, the peacock throws its wings open to reveal its spectacular, opulent upper wing surface patterns, with four large eyespots. It also makes noises. Eyespots are thus not a form of camouflage or mimicry but an extreme form of advertising. This isn't the warning coloration of toxic creatures, and the very many butterflies that have

---

* Just to confirm that eyespots really are functional, in 2005 Swedish researchers experimented with peacock butterflies, covering the eyespots of some. They found that 'only one of 34 butterflies with intact eyespots died during the 30-minute trials, whereas 13 of 20 butterflies with covered eyespots were killed'.

eyespots are not mimicking each other or anything else. Eyespots are not restricted to butterflies and moths: some mantises also have them.

The dry season forms of butterflies with eyespots are duller, with the eyespots much reduced, and leaf mimicry is common in the autumn. These are simple modifications of the groundplan, involving brightening or dulling of the pattern. For many butterflies, startle-with-bold-eyespots is the strategy for summer; camouflage-with-dead-leaf-forms is for the autumn campaign.

Different seasonal forms coming from the same butterfly (with exactly the same genes) are a bit like that harem of variably mimicking swallow-tails coming from the same brood of eggs. And it isn't only seasonal forms that can be created in the laboratory: geographical variants can also be produced by heat-shock and cold-shock. Applying cold-shock to the pupae of southern species sometimes produces forms very similar to more northern varieties. Goldschmidt reported that the European small tortoiseshell butterfly (*Vanessa urticae*), when cold-shocked, produces a form identical to a Scandinavian subspecies; and heat-shock applied to the latter produced a replica of the Sardinian variety. Here we have a north–south shuttle of varieties, induced by the simple trick of warming or cooling the pupae.

These apparently miraculous transformations suggest something remarkable about the genetic apparatus and the way it works in a developing animal. The experiments were early evidence for the subtle effects of the environment on genetics – the genes do not always produce the same effects in all conditions. This isn't Lamarckism – the inheritance by the next generation of characteristics acquired during life – but an example of the environment affecting the gene expression of a creature during the course of its development to an adult.

What have these seasonal forms got to do with mimicry? Well, the difference between the summer and the autumn form of a butterfly is quite a big leap: about as big as the jump from a non-mimicking butterfly to a mimic. And the seasonal forms occur without any mutation at all: simple changes of temperature (and light) lead to different forms of gene expression. For Goldschmidt, these transformations were powerful support for Punnett's idea that mimetic resemblances had evolved in one big leap rather than as a series of Darwinian small variations.

On this scenario, all the patterns are pre-existing and have only to be uncovered, through mechanisms as yet unknown. Goldschmidt described the choice of one seasonal form over another as being like a juke box whose arm selects one record or another. So his scenario for mimicry has one big genetic lever selecting a mimicking form from the genetic pattern book which he believed all butterflies to contain (the Nymphalid ground-plan). Similarly, he believed that the *Kallima* leaf plan could be derived from the standard butterfly groundplan in one big leap. (Suffert had only showed how the *Kallima* pattern could be derived geometrically from the groundplan – he did not propose a mechanism.) *Kallima* isn't a freak – many seasonal forms are dead leaf mimics – so this seemed plausible.

These heat-shock and cold-shock experiments were powerful and disturbing. To some they suggested the possibility of untapped powers in nature. Such ideas appealed to the planners of the Soviet Union, where a virulent new form of Lamarckism arose in the 1930s, under the aegis of the agronomist Trofim Lysenko (1898–1976). Lamarckism always seems rather plausible to the unscientifically trained (and even, as we have seen, to Darwin). It is obvious that a body can be developed by use during its life. So why shouldn't a man who turned himself into Mr Universe before he had children pass those muscles on? Unfortunately, it has been known since the late nineteenth century that every creature is born with the genes it is going to pass on, already formed. The environment has no way of influencing the germ line leading to the next generation – even though it can, as we have seen, influence the expression of genes after fertilisation has occurred, as in those seasonal forms.

The Soviet Union wanted to increase food production and its climate was severe. Lysenko managed to convince many people, including Stalin, that he could acclimatise plants to harsh regions and that the seeds sown from these vernalised plants, as he called them, would also be hardy. By this means winter wheat would grow when it was sown in spring.

These ideas became a major ideological battleground. Western genetics, which did not accept Lysenko's ideas, was denounced as bourgeois and reactionary. Soviet biology was ruined for decades and many biologists lost their careers – some even their lives. The Lysenko episode highlighted the dangers of knowing what answers we want from nature and refusing to listen to anything that might contradict them. As

Lysenko's colleague I. V. Michurin (1855–1935) put it: 'We cannot wait for favours from nature; we must wrest them from her.' But you cannot do it by falsifying results.

Lysenko was a real heretic – a wicked perverter of science – but in the 1930s Richard Goldschmidt was also treated as if he were beyond the pale. The 1930s was a decade of ideological clashes, of warring dogmas, and this mood also infected science. The prevailing mood in biology did not know what to make of the findings from the heat-shock and cold-shock experiments. As a scientist, Goldschmidt was in the line of de Vries and Bateson. He understood an aspect of genetics and evolution which has come to be generally accepted only in recent years, with the rise of Evolutionary Developmental Biology (Evo Devo): that genes work by controlling the rates of chemical processes. He also recognised that mutations at critical sites on a gene might have a large effect, similar to that of cold-shock.

Goldschmidt's ideas crystallised in the form of what he called 'the hopeful monster hypothesis'. This was asking for trouble, as he realised: 'I spoke half jokingly of the hopeful monster in my first publication on the subject, a lecture read by invitation in 1933 at the World's Fair in Chicago.' The hopeful monster was the provocative label given to a mutation which was: (a) large in effect; and (b) a positive genetic change which would confer advantages and would spread through the population – as opposed to the unviable large-scale 'Hopeless Monster' mutants collected by Bateson. The theory was an extension of Punnett's idea that a mimetic species came into being through a large mutation which more or less perfected the likeness, first time lucky. But Goldschmidt added his understanding of embryology – that is, of the processes of development. He realised that, while natural selection acts on the adult, allowing it to pass on its genes or not, any change in the organism that natural selection can act upon has to occur *in the embryo,* and alterations to the sensitive process of embryological development can produce large changes in the adult. These hopeful monster mutations are shifts in the timing of genes during development. Very often the forms produced will not be viable, but occasionally one will hit upon something new and interesting.

This is a refutation of the commonsense view that a genetic mistake in a complex creature must inevitably come to grief. Such an outcome is not necessarily the case, because developmental processes are not fixed, like a car production line. If in a production line something gets out of sequence, the process jams up; but, in the living development line, processes are often able to work round an error so as to compensate. Bateson's double-dactyly for instance – a human being with two hands joined at the thumb – is an example of a severe abnormality, but it is not life-threatening.

Goldschmidt thought that the delicate balance of developmental processes was a major constraint on evolution. If a mutation allowed the developmental process to proceed, a change was possible, but mutations that threw Cyclopean spanners in the works of that system would produce monsters. Goldschmidt gave examples of large evolutionary changes that could hardly have evolved by gradual steps. Bottom-dwelling flat-fish such as plaice and flounders have an uncanny mimetic ability to replicate the patterns of the sea floor they dwell upon; they also have another startling adaptation: they have both eyes on the same side. At some point in evolution, one eye has moved around from one side of the body to the other. Once there, it has an advantage but an eye that was merely half way would have no advantage at all. In fact, during embryological development the eye still physically travels around the fish to take up its new position.

Goldschmidt believed that the eye shift must have occurred as a hopeful monster leap in embryological development. In such a development, the eye would be very likely to be accompanied by the associated musculature and only minor mutations would be needed in future to perfect the new arrangement. In just the same way, Bateson had observed that, in monstrosities, when an organ moved to a new location, all of the component parts went with it.

Most biologists were hostile to Goldschmidt's ideas. The attacks on him became strident after he gave the Stillman Lectures at Yale in 1939 and published them as *The Material Basis of Evolution* (1940). In view of the response, Goldschmidt admitted that he 'certainly had struck a hornet's nest. The Neo-Darwinians reacted savagely.* This time I was not only crazy but almost a criminal.'

---

* Neo-Darwinianism, otherwise known as 'the modern synthesis', was the prevailing theory in biology. It harmonised Darwinian evolution with genetics and proposed that evolution had occurred by means of natural selection on small, cumulative genetic mutations. Prominent Neo-Darwinists were R. A. Fisher, Julian Huxley and E. B. Ford in Britain, and Theodosius Dobzhansky, Sewall Wright and Ernst Mayr in the USA.

The opposition to Goldschmidt's hopeful monster theory of large leaps in evolution was led by the English biologist R. A. Fisher (1890–1962). Fisher was the most distinguished mathematical geneticist of his generation, a firm believer in the Darwinian small incremental variation theory of evolution and an important figure in the English lineage of biologists working in mimicry – a line which extends from Bates, Wallace and Poulton, through Fisher, to present-day researchers. His 1930 book *The Genetical Theory of Natural Selection* was highly influential in setting the tone for decades.

Fisher's book has a chapter on mimicry which explains why the latter is such an important test for natural selection – a point Darwin had instantly realised on hearing of Bates's findings on mimicry in the Amazon. In most cases it is hard to know what factors in the environment are driving evolution, and equally hard to know which organs, patterns or physiological features are being acted upon by selection. But in mimicry, as Fisher stressed, we can be sure that predation is the environmental factor and that insect wing patterns and colours are the features acted upon. In science, the aim is always to find two linked factors whose interaction can be studied irrespective of what is going on around them. Where many causes conspire to produce complex effects, the wood can never be seen for the trees. Mimicry shows evolution in action in its most stripped-down, basic form.

As a Darwinian believer in evolution by means of small incremental changes, Fisher pointed out the weaknesses in Punnett's argument, which claimed that the fact that multiple mimetic females of swallowtail butterflies are controlled by a single gene indicates that they must have sprung into being by a sudden large leap. The variant female patterns of *Papilio* are polymorphic; so is the existence of sex. A single genetic switch also decides whether a human being will be male or female but no one is suggesting that women sprang from men in one single mutation (except the Book of Genesis, that is). The origin of sex is still unknown, but sex predated large creatures such as human beings by billions of years.

The contemporary English evolutionary biologist John Turner has put the case against the hopeful monster theory very cogently:

> Goldschmidt's argument that because major mutations occur very rarely, once in a while one would hit on just what was needed to make

a passable imitation of a leaf or another butterfly is like arguing that because our artillery has no problem in firing at a town ten miles away (in the course of the bombardment they will hit some houses), there's no difficulty in hitting the house at 55 High Street.

Fisher went further: he was sure that large leaps in a gene would, almost certainly, always produce effects not just in the desired part of the body, but in many other parts, and these would lead to a creature being so unfit that it would not survive and breed. End of story.

To see what the critics of the hopeful monster hypothesis were getting at, consider the cyclopic sheep of Utah. There, in the 1950s, 5–7 per cent of sheep were born with a usually lethal condition known as cyclopia. Like the legendary monster of Homer's *Odyssey*, these sheep had only one eye. This feature was associated with other deformities: like the eyes, the brain hemispheres did not divide – the head, in fact, was a complete mess. It was eventually discovered that the cause was a lily plant, *Veratrum californicum*. This plant contains a chemical (now called cyclopamine) which, if ingested between days ten to fourteen of foetal life, disrupts the embryo's development in this systematic way. The result of this spanner in the works is multiple deformities, because gene action has been altered at many different stages of development by a single mutagenic chemical.

The quarrel between the Darwinian small, cumulative variations camp and the large, discontinuous leaps school is reminiscent of the two sects in Jonathan Swift's *Gulliver's Travels*, the Little-Endians and the Big-Endians. The object of contention there was the breakfast boiled egg: which end to crack? Swift was not thinking of scientific controversy when he wrote this, but more of the way religions or political movements tend to splinter into factions over just such tiny doctrinal points. But the need to pick a bone of contention is such a deeply ingrained property of the human mind that, inevitably, it has also permeated science.

There has always been excellent evidence for at least one kind of hopeful monster. The skeleton of a snake is just one long backbone: they have between 130 and 400 vertebrae and, in evolving, extra vertebrae have been added on many occasions. This process cannot have occurred by fractional increases (a bit of a vertebra would jam the spinal column, rendering the snake a 'hopeless monster'). Even one of Goldschmidt's fiercest later critics,

Richard Dawkins, admits that, 'in the evolution of snakes, numbers of vertebrae changed in whole numbers rather than fractions . . . *It is easy to believe that individual snakes with half a dozen more vertebrae than their parents could have arisen in a single mutational step* [my italics].'

But a re-examination of Goldschmidt's ideas was to begin in 1977, with Stephen Jay Gould's essay 'Return of the Hopeful Monster', which opens:

> Big Brother, the tyrant of George Orwell's *1984*, directed his daily Two Minutes Hate against Emmanuel Goldstein, enemy of the people. When I studied evolutionary biology in graduate school during the mid-1960s, official rebuke and derision focused upon Richard Goldschmidt, a famous geneticist who, we were told, had gone astray.

But Jay Gould concluded:

> In my own, strongly biased opinion, the problem of reconciling evident discontinuity in macroevolution with Darwinism is largely solved by the observation that small changes early in embryology accumulate through growth to yield profound differences among adults.

Goldschmidt's observation of the fluidity of gene action led him to doubt that genes were particles at all – again, an idea which was way ahead of its time. Following Mendel's work on the factors of inheritance, exhaustive breeding experiments on fruit flies had indicated that specific genes were located at definite positions on the chromosomes. But Goldschmidt thought the genes were more plastic than that: he envisaged a 'reshuffling or scrambling of the intimate chromosomal architecture'. This was astonishingly prophetic, anticipating the existence of what we now call mobile genetic elements.* Goldschmidt was so much swimming against the tide that the value of his work would not be realised for fifty years.

---

* Evidence began to emerge in the 1940s to challenge the abstract view of the gene. The American geneticist Barbara McClintock's (1902–92) experiments with maize showed many mutations occurring within parts of the same plant, which suggested that genes were more mobile than the modern synthesis could account for.

CHAPTER 9

THE NATURAL HISTORY OF THE VISUAL PUN

If all of these interpretations were being done consciously we would be tempted to call it art. Or perhaps a set of brilliant puns. Perhaps both. Certainly the closest analogy to whatever process it was that led the angler fish to use an extended fin-spike as a luminous bait, is what was going on when Marcel Duchamp hung a lavatory seat in an art gallery.

Richard Mabey, *In a Green Shade*

If Thayer was the artist who had the greatest impact on our knowledge of camouflage in the natural world, there was no doubt who was the writer of mimicry: the Russian novelist Vladimir Nabokov (1899–1977). Nabokov's passion for butterflies began early: 'From the age of seven, everything I felt in connection with a rectangle of framed sunlight was dominated by a single passion . . .' This passion was both the traditional ambition of the lepidopterist – he longed to have a new butterfly species named after him – and something more.

Nabokov brought a hyper-refined artist's eye to butterflies, and the conflicts which emerged between him and mainstream biologists make a fascinating case study in the story of science and art: the so-called 'two cultures'. Nabokov's science was the old-fashioned one of identifying and describing species in minute detail and classifying them according to their similarities and divergences. He became an expert in the minutiae of tiny anatomical distinctions, especially of the genitalia (all butterfly species have distinctive genitalia, which function as a lock and key; only members of the same species can mate, because only they fit).

In his 'butterfly' novel *The Gift*, written in Russian between 1935 and 1937 but not published in English until 1962, he refers to heat-shock and cold-shock, but takes them no further than appearances:

> He variously warmed and cooled the golden chrysalids of my tortoise-shells so that I was able to get from them Corsican, Arctic, and entirely unusual forms looking as if they had been dipped in tar and had silky fuzz sticking to them.

He was especially drawn to mimicry and described its appeal in wonderfully chiselled prose. Nabokov enlarges our appreciation of mimicry by showing that butterflies are not just beautiful in the conventional sense: the whole system of resemblances, often to such fantastic objects as dung, is artistic. A caterpillar or spider that perfects a resemblance to bird-dung is like a Dutch still-life artist lovingly recreating the texture of folded velvet.

But Nabokov fell into the trap of quarreling with Darwinian science an account of his artistic interest in mimicry. *The Gift* (1937) is a novel, but the protagonist, like Nabokov, is a lepidopterist. Nabokov cannot resist making him a mouthpiece for his views:

> He [the protagonist's father] told me about the incredible artistic wit of mimetic disguise, which was not explainable by the struggle for existence (the rough haste of evolution's unskilled forces), was too refined for the mere deceiving of accidental predators, feathered, scaled and otherwise (not very fastidious, but then not too fond of butterflies), and seemed to have been invented by some waggish artist precisely for the intelligent eyes of man.

Nabokov's anti-Darwinian attitude is puzzling and it would be easy to account for it by dismissing him as a mere amateur butterfly hunter. But Steven Jay Gould, who worked twenty years later in the same institution (Nabokov was employed at Harvard from 1942 to 1948 as research fellow and unofficial curator in Lepidoptera at the Museum of

Comparative Zoology), stressed that Nabokov was 'a fully qualified, clearly talented, duly employed professional taxonomist, with recognised "world-class" expertise in the biology and classification of a major group'.

For Nabokov, as for a few others before him, the degree of imitation found in natural mimicry was simply too extreme to allow that it should have come about through natural selection, though he had no clearly formulated alternative. Strangely, when he wanted to emphasise the unlikelihood of natural selection achieving the miracles of mimicry, he chose some dubious examples. In *The Gift*, he refers to

> Chinese rhubarb whose root bears an uncanny resemblance to a cater-
> pillar . . . while I, in the meantime, found under a stone the caterpillar
> of an unknown moth, which represented not in a general way but with
> absolute concreteness a copy of that root, so that it was not clear which
> was impersonating which and why.

This passage has been expertly decoded by the German Nabokov specialist Dieter Zimmer. Nabokov had taken this example not from life but from a book by A. E. Pratt, *To the Snows of Tibet through China*, 1892. Pratt had written: 'The medicines collected here are rhubarb, *Tchöng tsaö* (*Sphaeria sinensis*), a plant the root of which bears an almost exact resemblance to the body of a caterpillar, and *pey-mou*.' As Zimmer realised, Nabokov had taken the name *Tchöng tsaö* to be a Chinese name for rhubarb, when in fact Pratt's sentence lists three separate plants: (1) rhubarb; (2) *Tchöng tsaö*; (3) *pey-mou*. To compound the error, Pratt had been wrong in thinking that the plant was mimicking a caterpillar: Zimmer discovered that *Tchöng tsaö* is a fungus which grows on the caterpillars of a ghost moth, devours them and grows out of their carcass. As Zimmer says, the fungus is the caterpillar (or what remains of it).

This episode highlights a danger for those who are as enamored of mimicry as Nabokov was: the trap Poulton fell into – the danger of seeing imitation everywhere. Zimmer describes two more such cases in *The Gift*. Nabokov's character refers to a 'cunning butterfly in the Brazilian forest which imitates the whir of a local bird'. This is easily traced to Bates's book *The Naturalist on the River Amazons*, which Nabokov admired. Bates describes the resemblance between the hummingbird hawkmoth and hummingbirds, saying: 'Several times I shot by mistake a hummingbird hawk-moth instead of a bird.' He goes on to discuss the resemblance, which the natives find uncanny, but Bates attributes this to the identical lifestyles of these creatures, which hover before flowers to sup nectar. Neither Bates nor any other reputable authority has ever claimed this behaviour to be mimicry.

Of the final example cited in *The Gift*, a caterpillar imitating a yellow flower, Zimmer concluded: 'After studying many botanical and entomological species and genus catalogs from 250 years I felt safe enough to conclude that Nabokov had invented both the moth (*Pseudodemas tschumarae*) and its food plant (*Tschumara vitimensis*).'*

Another of Nabokov's heresies, one he repeatedly returned to, is the idea that the display of mimicry was somehow *intended for our eyes*. *The Gift*, as it is now published, includes his Addendum, 'Father's Butterflies', which is entirely devoted to butterflies and mimicry. Nabokov's hypersensitivity and his delight in the theatre of form and gesture that mimicry creates rendered him incapable of believing that the show could have existed for millions of years without an observer able to appreciate it – an extremely narcissistic attitude. He spoke of

> . . . the fantastic refinement of 'protective mimicry', which, in a world lacking an appointed observer endowed with artistic sensitivity, imagination, and humor, would simply be useless (*lost upon the world*), like a small volume of Shakespeare lying open in the dust of a boundless desert.

And a little later:

> Yet, long before the dawn of mankind, nature had already erected stage sets in expectation of future applause, the chrysalis of the Plum *Thecla* [*Strymonidia pruni*, the Black Hairstreak] was already made up to look like bird droppings, the whole play, performed nowadays

---

* To be fair to Nabokov, there are caterpillars that imitate yellow flowers: the caterpillars of camouflaged looper moths of the genus *Synchlora* snip yellow petals off with their mandibles and stick them onto their back.

with such subtle perfection, had been readied for production, only awaited the sitting down of the foreseen and inevitable spectator, our intelligence today.

Even this does not exhaust his variations on the theme: a rather vulgar idea for such a fastidious writer. It is the Bible that says that the creatures were made for Man's use and pleasure, and it does not take scientific genius to realise that such an idea cannot be part of our study of nature. Very many species perished before humans were able to observe them – and some of them might have been accomplished mimics.

Nabokov makes much of the fact that the panoply of mimicry seems complete in our time: 'In nature as it exists today one does not note forms of half- or quarter-resemblance.' This is not true: cases of imperfect mimicry abound (see p. 242).

We shall return to the questions Nabokov raises, but the unravelling of the mystery of mimicry was not his concern: what he contributed was a loving evocation of the glories of the phenomenon, his writing matching the swaggering display of the creatures themselves:

> The mysteries of mimicry had a special attraction for me. Its phenomena showed an artistic perfection usually associated with man-wrought things. Consider the imitation of oozing poison by bubblelike macules on a wing (complete with pseudo-refraction) or by glossy yellow knobs on a chrysalis ('Don't eat me – I have already been squashed, sampled and rejected'). Consider the tricks of an acrobatic caterpillar (of the Lobster moth)* which in infancy looks like birds' dung, but after molting develops scrabbly hymenopteroid appendages and baroque characteristics, allowing the extraordinary fellow to play two parts at once (like the actor in Oriental shows who becomes a pair of intertwisted wrestlers): that of a writhing larva and that of a big ant seemingly harrowing it . . . 'Natural selection', in the Darwinian sense, could not explain the miraculous coincidence of imitative aspect and imitative behaviour, nor could one appeal to the theory of 'the struggle

---

* This is Poulton's favourite moth again, the multiple mimicry of which was so mocked by McAtee.

for life' when a protective device was carried to a point of mimetic subtlety, exuberance, and luxury far in excess of a predator's power of appreciation. I discovered in nature the nonutilitarian delights that I sought in art. Both were forms of magic, both were a game of intricate enchantment and deception.

Nabokov points up the art–science divide in a far more interesting way than the professional/amateur dichotomy. As Gould stressed, what Nabokov brought to both science and art was an inordinate respect and affection for detail: 'In high art and pure science, detail is everything.' This begs the question in both departments. Detail is important in both art and science, but it is not enough for either. In Nabokov's writing, the burnished details are at the service of an orchestrated work of art, with character, drama and plot, but in biology he seems to have been totally uninterested in the big picture: the genetics and the evolutionary history of his butterflies concerned him not all. He was a good taxonomist because his passion for detail led him to spend countless eyesight-destroying hours at the microscope, unravelling the minute tell-tale differ-ences in genitalia that distinguished one butterfly species from another.

But Nabokov's passion for observation fed his disdain for deep biology. As Gould commented: 'Nabokov frequently stated that his non-Darwinian interpretation of mimicry flowed directly from his literary attitude – as he tried to find in nature "the nonutilitarian delights that I sought in art." ' A scientist such as Darwin or Richard Goldschmidt is not trying to do any such thing. Nabokov's aesthetic approach is comple-mentary to science – it humanises it and allows it to become, in Wordsworth's and Coleridge's words, 'a dear and genuine inmate of the household of man' – but it must not interfere with the work of doing real science. But when, in the novel *Ada*, he has a character say – 'how incan-descently, how incestuously – *c'est le mot* – art and science meet in an insect' – yes, we feel, this is the true meeting place: the thrill must not be driven out of science. Few scientists have ever described passionate absorption in scientific work as Nabokov has:

The tactile delights of precise delineation, the silent paradise of the camera lucida, and the precision of poetry in taxonomic description represent the artistic side of the thrill which accumulation of new knowledge, absolutely useless to the layman, gives its first begetter . . . There is no science without fancy, and no art without facts.

Nabokov was a precisian, a literary stylist and a stickler as an entomologist, but he was out of step with the intellectual currents of his time. He had the permanent émigré's nostalgia for the lost culture. And his penchant for deception was integral to his greatest novel: *Lolita* has a man wooing and marrying a woman just to get close to the real object of his desire – her daughter.

But mimicry appeals to more than one kind of artist. For Nabokov it is the exquisite precision of naturalistic copying that chimed with his ornate, hyper-realist prose style. For him, nature was another finicky translator of patterns from one realm to another, just as he translated idioms and wordplay from Russian into English. For the French writer and *maître à pensée* Roger Caillois (1913–78), nature was a surrealist. Surrealism would come to have an intimate connection with mimicry and camouflage, and it is most apparent in the work of Caillois.

Caillois joined the surrealist movement in 1932 and his first writings on mimicry appeared in 1934, in his essay 'La Mante religieuse' ('The religious mantis'). Surrealism is always associated with lobsters, thanks to Salvador Dali's sculpture *Lobster Telephone*, a real black telephone with a plaster lobster sitting on the cradle; but perhaps the praying mantis has a better claim to be the surrealist pet? André Breton, Paul Eluard and Salvador Dali, besides Caillois, were all fascinated by mantises; Eluard had a collection and Breton bred them for two years. Why? Caillois said: 'Certain objects and images are endowed with a comparatively high degree of lyrical force because their form or content is especially significant.'

The mantis is obviously one such. Caillois cites its 'remarkably anthropomorphic appearance'. In fact, it isn't exactly anthropomorphic: with its triangular head, long lunging arms and strange startle dances, it is the perfect cartoon alien monster. Mantises look like cartoon insects; they look as if they had been designed on the *Bug's Life* computer.

Caillois's attraction to mimicry seems to be a fixation with a secret world, non-human, with rules of its own:

Finally, let us not forget the mimicry of mantises, which illustrates, sometimes hauntingly, the human desire to recover its original insensate condition, a desire comparable to the pantheistic idea of becoming one with nature.

He follows this with examples of mantis mimicry, including the orchid mimicry, first noticed by Annandale in the 1890s: '*Hymenopus bicornis*, which is hard to distinguish from a simple, marvelous orchid'.

In 1935 Caillois wrote an essay called 'Mimicry and legendary psychoasthenia'. The surrealists were interested in states that belied the commonsense distinctions – conscious/unconscious, asleep/awake, the individual/the environment – and Caillois takes as his starting point the idea that mimicry is a pathology of the normal distinction between a creature and the environment. Caillois had a good knowledge of the range of mimicry in nature and was not afraid of disputing the zoologists' interpretations. He doubted the biological efficacy of mimicry:

> Predators are not at all deceived by homomorphy or homochromy:* they eat acridians blended into the foliage of oak trees, or weevils resembling tiny pebbles, which are quite invisible to man's naked eye. The phasmid *Carausius morosus* (which uses its shape, color, and posture to simulate a plant twig) cannot be kept out in the open because sparrows immediately discover and devour it.

Like Nabokov, Caillois believed that the phenomenon of mimicry had a meaning way beyond the struggle for life between predators and prey: 'The objective phenomenon is the fascination itself. This is illustrated, in particular, by the *Smerinthus ocellata*,† which does not look like anything dangerous at all. Only the eye-shaped markings come into play.' Caillois saw mimicry in nature as essentially an image-making process which anticipated the work of the human artist:

---

* Homomorphy means 'of a similar pattern'; homochromy, 'of the same colour'.
† The eyed hawkmoth, which has rose-tinted hindwings with dramatic, large, blue and black eyespots.

Morphological mimicry could then be genuine photography, in the manner of chromatic mimicry, but photography of shape and relief, on the order of objects and not of images; a three-dimensional reproduction with volume and depth: sculpture photography.

Scientifically, this doesn't make any sense – the process by which mimicry has evolved bears no relation to photography – but what he was groping for is the fact that the pattern of one object – say, a dead leaf – has been mapped onto another: a *Kallima* butterfly. Caillois does not even mention natural selection. He haughtily avoids the subject. What interests him is that certain creatures are not what they are: *Kallima* is a butterfly, but, to certain other creatures at other times, a leaf. He is interested in human analogues of this dual nature. For him, mimicry is part of a world of signs, signals, patterns and deceptions: a parallel universe to this human one. He has an urge to escape into this realm, quoting the conclusion of Flaubert's *The Temptation of Saint Antony*, in which the protagonist ceases to be human:

> I should like to have wings, a shell, bark, to breathe out smoke, to sport a trunk, to wrench my body, to split myself everywhere, to be everywhere, to emanate perfumes, to grow like a plant, to flow like water, to vibrate like sound, to shine like light, to squat on every form, to penetrate each atom, to descend to the bottom of matter, to be matter itself.

In Flaubert, the prelude to this desire is a fantasy of mimicry in which 'plants are now no longer distinguished from animals . . . Insects identical with rose petals adorn a bush . . . And then plants are confused with stones. Rocks look like brains, stalactites like breasts . . .' Where does the surrealism come in? Caillois finds that, in some of Salvador Dali's paintings from around 1930, 'Whatever the artist may say, these men, sleeping women, horses, and lions . . . result less from paranoid ambiguities and multiple meanings than from the mimetic assimilation of animate beings into the inanimate realm.'

Surrealism was many things to many people, but it's safe to say that in surrealist paintings things are not themselves. Odd juxtapositions, things out of context, gross distortions of natural forms and a blurring of the distinction between living and non-living things were at the heart of the movement. The surrealists adopted as a slogan a line from the nineteenth-century French poet, the self-styled Comte de Lautréamont (1846–70), who defined beauty as 'the chance meeting of a sewing machine and an umbrella on a dissecting table'.

Surrealism's essential link with mimicry is that blurring of the line between living and non-living. In Salvador Dali's *The Persistence of Memory*, a clock wraps itself over the edge of a table and time drips away like an amoeboid jelly. In René Magritte's *The Pleasure Principle* (a title presumable inspired by Freud's concept), a man seated at a desk has, instead of a head, a large light bulb glow (sans the bulb) with diffused corona. In Magritte, visual and verbal meanings play off against each other. In *The Use of Words 1*, a gentleman's pipe rendered in advertising air-brush style sports the caption '*Ceci n'est-ce pas une pipe*' ('This is not a pipe'). Creatures in mimicry are in effect saying: 'I am not what I am.'

Max Ernst's 1921 painting *Celebes* features a machine/animal hybrid. Ernst, a native of Cologne who fought in the First World War, was a pioneer surrealist who began in the related Dada movement, in Zurich, in 1916. The Dadaists countered the cruel absurdity of the war with harmless artistic absurdities. The *Celebes* is a man-made form that strikes an eerie note in most viewers. What is especially striking now is that it appears to be mimicking a cylinder vacuum cleaner of the type manufactured by Henry, something that could not possibly have been in Ernst's mind.* The form is actually based on a large corn bin of the Konkomwa tribe in Sudan. What is fascinating is its mimetic human form. It has two stout legs merging into a pot belly, like a fat trencherman. The surrealists acknowledged Lewis Carroll's nonsense verse as a precursor, and it isn't surprising that the name 'Celebes' comes from a scurrilous schoolboy rhyme: 'The elephant from Celebes / Has sticky, yellow bottom grease.'

Ernst characterised the painting as 'the systematic exploitation of the coincidental or artificially provoked encounter of two or more unrelated

---

* The cylinder vacuum cleaners manufactured by the Henry company have a particularly anthropomorphic appearance. The current range has two eyes painted on to create a face, with the nozzle as the nose.

realities on an apparently inappropriate plane and the spark of poetry created by the proximity of these realities' (de Lautréamont's principle again). Time has amplified the poetry by inventing the cylinder vacuum cleaner. This 'artificially provoked encounter' is exactly what cartoonists employ when they link the visual and the verbal world in puns which use visual images to amplify verbal messages. As Gombrich was aware, the system of signs and gestures signalling intentions in the human sphere and in nature shows some intriguing parallels, from the simple use of red, yellow, black and white warning signs to the obscure symbolism of surrealist art.

Caillois's thinking is typical of the advanced aesthetic theorising of the period. What is interesting is that writers of his persuasion were drawn towards mimicry at a time when blurring the distinction between the individual and the environment was shortly to become more than a game for the art salon. Artists would shortly find themselves at war again, and some of them would be drawn to camouflage. What cubism was to the First World War, surrealism would be to the Second.

The old order was coming to an end. In the summer of 1939 the English surrealist painter Roland Penrose (1900–84) visited Picasso on his Mediterranean stamping ground. Oblivious to the impending disaster, Picasso was painting *Night Fishing at Antibes*, one of his most evocative paintings. It applies the cut-out grotesqueries that he employed in *Guernica* to an eroticised night scene – the essence of Mediterranean mystery and sensuality. Penrose was in Antibes when the news of Germany's invasion of Poland came through. He and many others set off for home. In his autobiographical *Scrap Book* he says:

> On the boat I found Julian Trevelyan* also on his way home and, wondering how either of us could be of any use in an occupation so completely foreign to us both as fighting a war, we decided that perhaps our knowledge of painting should find some application in camouflage.

* Julian Trevelyan (1910–88) came from a noted English family: his father was a poetaster on the fringes of Bloomsbury and his uncle was the historian G. M. Trevelyan. Julian read English at Cambridge and moved to Paris to become a painter. His work was included in the International Surrealist Exhibition (1936), which launched surrealism in the British consciousness.

CHAPTER 10

CANNIBALS AND SUNSHIELDS

Yet, while there is the closest analogy between the needs for concealment and deception in nature and in war, in this sphere the coloration of animals has attained a degree of perfection far beyond the comparatively crude attempts at camouflage with which we are too easily satisfied – attempts which often neglect to make use of the very principles revealed in the coloration of innumerable snakes, caterpillars, birds, fishes and other organisms.

Hugh B. Cott, *Adaptive Coloration in Animals* (1940)

The coming of war goaded the surrealist painters Roland Penrose and Julian Trevelyan into taking up camouflage as an appropriate form of war work. And for biologists, the war would once again pose the question: can the principles of camouflage in nature be applied to human combat? Naval warfare appeared pretty much the same in 1939 as it had been in 1918; but, as the Second World War loomed, one overwhelming change was in prospect. Aerial warfare had begun in the middle of the First World War: it was widely expected to dominate the new conflict. The slogan on everybody's lips was: 'The bomber will always get through.'

The new dominance of air power had strong implications for camouflage. Aeroplanes had three functions in war: to attack ground installations by bombing; to achieve superiority in the air by destroying the enemy's fighter aircraft; and to mount reconnaissance missions, bringing back detailed photographs of potential targets. Although in air combat speed, manoeuvrability and the skill of the pilot were paramount, in Britain and Germany combat aircraft themselves sported simple camouflage. The upper surfaces of British planes were painted in bold green-and-brown disruptive patterns, the underside in pale blue. The value of the disruptive effects of the upper surfaces was dubious. Seen from above, the motion of the plane rendered disruption ineffective. German aeroplane camouflage seemed to be based on the idea that they would always stay at a higher altitude and be viewed from below: the underside was pale blue, but the upper parts were also blue, with mottled and stippled grey and brown markings, designed to break up the outline when they were seen against sky and cloud, but not the land. American aircraft began the war with all-over olive drab coloration, but, with their increasing dominance in the air, all the American fighter planes were left unpainted, with a natural silver aluminium finish.

Ground installations such as airfields, factories and concentrations of troops, guns and supplies were highly vulnerable to attack from the air and posed a major challenge for the practitioners of camouflage. The landscape looks very different from the air, and the principles of camouflage which had served in ground combat might not be appropriate in the new era. In the 1930s John Graham Kerr was still smarting from the rejection of his camouflage ideas by the Admiralty in the First World War, but he fully understood the new threat and the opportunity posed by aerial warfare. This time he had a secret weapon: his acolyte Hugh Cott, who had become Britain's leading expert in mimicry and camouflage in nature.

In 1935, Kerr had resigned his university post as professor of zoology at Glasgow to devote himself to the question of camouflage in the coming war. As MP for the Scottish universities, he pestered every authority he could think of in government and the services to have Cott installed as camouflage supremo. He was hampered by apathy, deep interservice rivalry, and by his own knack of rubbing people up the wrong way. But Kerr was not completely without influence: in June 1939 he was knighted 'for political and public services in Scotland', and as an MP he could now ask questions in Parliament.

Hugh Bamford Cott (1900–87) was a son of the manse and, after an education at Rugby and Sandhurst, he served in the Army in Ireland during the troubles of 1919–21, where he first became aware of military camouflage. He went up to Cambridge in 1922 to study theology, but an undergraduate trip to Brazil in 1923 quickened his interest in wildlife and he switched to natural science on his return. A further trip to Brazil (1925–6), which included visiting Bates's terrain near Para, made a deep impression on him with its countless examples of mimicry, warning coloration and camouflage. He echoed Poulton on the connection between the fertility of the tropics and the existence of mimicry:

> And it is in the tropics especially, where there is such wealth of life and where competition is correspondingly acute, that these methods of offence and defence evolved by insects and other groups present problems of peculiar interest.

In Brazil he saw Bates's *Heliconius* butterflies: 'The heliconids are conspicuous, gliding in leisurely fashion, as if fully aware of their nauseous qualities.'

Cott's work as a lecturer in zoology at Glasgow University from 1932 on brought him into contact with Kerr. They shared a deep interest in mimicry and camouflage, and Cott believed in the applications of biological camouflage to warfare just as fervently as Kerr.

In the 1930s, Cott combined fieldwork with amassing a vast body of knowledge on camouflage and mimicry in the natural world. In the course of this he was involved in a long tussle with Waldo Lee McAtee, the American ornithologist who had crossed swords with Poulton in the Edwardian era. In 1933, McAtee unwisely took issue with a paper by Cott about poisonous East African tree frogs and their prey. So opposed was McAtee to the theories of protective coloration that he denied that some frogs exhibit warning coloration. Even when Cott declared: 'Poisonous secretions are of common occurrence among [frogs]', McAtee demurred. In his riposte, Cott listed some of the toxic amphibians. Many are very poisonous indeed and advertise the fact with remarkably lurid colouring – for instance the *Dendrobates* species, which come in every bright hue, from an eerie, glistening deep blue to black-spotted yellow and black-spotted red, and are famously used by indigenous peoples as arrowtip poisons. Despite this, McAtee could still blithely utter: 'There is no evidence of any frog being dangerously poisonous either to snake or bird predators upon it.'

Cott's demolition at this point was conclusive. And yet . . . McAtee disliked the smug consensus among zoologists, especially British zoologists, who assumed that natural selection drives evolution and then interpreted everything in terms of this assumption. As Gerald Thayer had in 1918, Cott was not above using circular reasoning to bolster his case:

> Dr McAtee is apparently either unable or unwilling to appreciate the function of a disruptive colour scheme, which may – as every field naturalist knows – be most effective in rendering more difficult the recognition of an animal by its enemies. It is easy for Dr McAtee to deny the effectiveness of this type of camouflage; but the experience of the Great War proved that it was neither easy nor expedient to dispense with precisely this principle, which was applied in the so-called 'dazzle' painting of ships with conspicuous success as a means of defence against submarine attack.

Thayer and Kerr had argued *from* nature that disruptive colouring could be useful when applied to ships. Cott was now importing this argument back *into* nature, to demonstrate the efficacy of disruptive colour. So dazzle was justified by nature and nature was justified by dazzle. In fact, Cott's argument from military dazzle camouflage was highly dubious, because the efficacy of the latter had hardly been proven, as we saw in Chapter 6; the Royal Commission on Awards to Inventors, commenting: 'The effect of "Dazzle Painting" as a protection against submarine attack proved to be disappointing in practice.' All the old arguments from the First World War were about to be rehashed.

In 1938 Cott moved from Glasgow to Cambridge as curator of invertebrates at the Museum of Zoology and university lecturer. As Kerr lobbied to have Cott installed at the heart of military camouflage, Cott was writing his magisterial synthesis of all that was known on camouflage and mimicry in nature, to be published as *Adaptive Coloration in Animals* (1940). Despite Kerr's prickliness, Cott's commonsense and commanding knowledge of natural camouflage gradually won him influence and the chance to show what he could do. In 1938 he gave several talks to military institutions and paid a visit to Camberley Staff College in March which was particularly successful; Cott reported: 'I am now regarded as an authority on camouflage by these people.' 'These people' included men from the Royal Aircraft Establishment, Farnborough (RAE), home of RAF research and development.

Cott was not wrong in his belief: in April 1939 he was given the chance to demonstrate his principles by camouflaging an RAF station at Mildenhall in Suffolk. As he began, he was aware there was a tussle brewing. Kerr and Cott learned that Norman Wilkinson, Kerr's adversary in the First World War, was being considered as camouflage supremo by the RAF. During that spring and summer, as a prelude to the new war, the old battles over dazzle painting in the First World War were refought in the pages of *The Times*.

The confrontation began with a long article by Cott in which, once again, he made the case for Kerr's priority in the use of disruptive patterns on the ships. Wilkinson duly responded with his trump card:

> Since Kerr maintains he was responsible for 'dazzle' may I point out that among the claims for 'dazzle' painting submitted to the Royal Commission of Awards for Inventors of which Mr. Graham Kerr's was one, and where the matter was thoroughly thrashed out, my claim was the only which received an award.

Cott tried to defend his friend by asserting that Wilkinson's 'dazzle' and Kerr's disruptive patterning were optically the same thing. But, while the bitterness of the old quarrel was played out again, the new struggle was over what kind of camouflage there would be in the coming war. Cott had ended his original article by claiming that camouflage is 'a child suffering from arrested development'. Wilkinson retorted: 'the child has turned the corner and is doing quite well.' These brief remarks would characterise the two approaches once war began: Cott – earnest, anxious and bustling; Wilkinson – laid back to the point of complacency. Wilkinson's role at the Air Ministry became formalised when the war broke out. Wilkinson and Cott were set on a collision course.

The question of the protection of airfields was urgent. They were highly conspicuous and, instead of attempts to hide such large installations, the authorities proposed the idea of dummy airfields with dummy planes from 1938, with the aim of encouraging the enemy to waste his bombs on decoys. The Air Ministry decided that the most likely people to rustle up dummy planes were those habitual stagers of artificial reality – the film studios.

The link between the film industry – the masters of illusion – and the war effort had in fact begun even earlier. In January 1937, the Hungarian refugee film-maker Alexander Korda, director of *The Private Lives of Henry VIII* (1933), who was to make *The Third Man* later on, in 1949, had offered his Denham Studios to the military, more or less with *carte blanche*:

> The staff employees, buildings and workshops are admirably suited to form a Camouflage Unit, if such a unit is required. These men are specialists at 'make believe' and deception in defeating both the eye and the camera. They possess the workmen, material and shops to build jerry constructions for deception purposes . . . Even their 2000 lights might be used for deception in floodlighting empty fields to resemble aerodromes.

A Fighter Command officer, the splendidly titled Captain and Brevet-Major P. G. Calvert-Jones, who went to check on the potential of this offer, was impressed by Korda's patriotic ardour, but didn't see how a civilian workshop could become a military installation overnight.

Korda's well-meaning civilian initiative was not the only one of its kind. A couple of years later, the surrealists were getting in on the act. True to their vow made on the boat home to England in 1939, the surrealist artists Roland Penrose, Julian Trevelyan and a few colleagues started a commercial company, the Industrial Camouflage Research Unit, soon after the outbreak of war in September 1939. The émigré architect Ernö Goldfinger* lent them his offices at 7 Bedford Square, London. The brief life of this unit (it closed in June 1940) was a cameo of progressive artists in action.

Julian Trevelyan described the casual amateurism of their efforts in his memoir, *Indigo Days*:

---

* Ernö Goldfinger (1902–87) was a modernist Hungarian architect who lived in England from 1934. After the war he built many tower blocks, both offices and housing.

Already the rash of squiggly green patterns was breaking out in every part of the country . . . At the Industrial Camouflage Unit . . . we were soon busy making models and painting them with abstract patterns so as to merge them, so we fondly hoped, into the background . . . In these early days of the war Industrial Camouflage was a perfunctory ritual that had very little to do with its proper function, which was, presumably, to protect targets from being bombed by enemy aircraft. Much of it was done in an uncoordinated freelance way; the little corner garage in a housing estate was quickly painted in wiggly lines, and so was the laundry and even the roof of the cinema. People seemed to feel that the green stripes were a charm that somehow brought them immunity from the unknown hazards of war.

The artists in effect enjoyed themselves, seeing wartime camouflage as painting by other means. They did not know what camouflaged buildings looked like when seen in air reconnaissance pictures. Were they camouflaging against day or night bombers? Despite the amateurism of the unit, the Player's Imperial Tobacco Company in Bristol commissioned them to produce 'a scheme that was in effect a huge abstract picture painted over the roofs and chimneys of an industrial town'. The work was paid for, but never used.

Goldfinger, who abhorred slovenly behaviour, fell out with the artists. They were late payers and abused his equipment. Goldfinger later confessed that the unit was 'the biggest con I was ever involved in'. The unit gained no more commissions and quickly broke up. The government forbade commercial companies to operate in war camouflage, and the protagonists were absorbed into the official world of military camouflage.

But meanwhile, in another neck of the woods, Mildenhall, Suffolk, Hugh Cott had an airfield to camouflage. He described his progress in letters to Kerr. An airfield is a big target: 'At Mildenhall the hangars are glaringly conspicuous, and have a serrated roof-structure which catch [sic] the eye from a great distance.' Cott regarded flying as absolutely necessary for his work and took to the air as often as possible.

It is not clear how much of the airfield Cott camouflaged – his letters to Kerr only mention a bomb store, which presented a strong squared pattern in oblique sunlight. For this, Cott devised a net held by concrete

1 Some of Henry Walter Bates's mimetic butterflies. The top row are '*whites*' (*Leptalis*), which mimic the *Ithomias* in the next row. The middle 'white' demonstrates how far the mimics have diverged from the typical form. Row 3 shows *Leptalis* mimicking other forms of Heliconidae (Row 4).

2 A Dennis-rayed postman butterfly (*Heliconius melpomene*) and its passion flower host plant.

3 The pioneers of mimicry and its role in evolution, against a rainforest background: (top to bottom) Henry Walter Bates, Alfred Russel Wallace and Charles Darwin.

4 The Malaysian orchid mantis eating a fly lured by the orchid.

5 The 'snake's-head' warning pose of the spicebush swallowtail caterpillar, *Papilio troilus*.

♀ PARENT.

♀ form hippocoon.

DANAINE MODELS.

OFFSPRING.

Amauris niavius dominicanus ♀.

♀ form hippocoon.

Danaida (Limnas) chrysippus ♂.

♀ form trophonius.

Amauris albimaculata ♀.

♀ form cenea.

Amauris echeria ♂.

♀ f. cenea.

6 Four different mimicking female forms of the swallowtail butterfly *Papilio dardanus*, bred from a single brood. The parent is the lone butterfly at the top; the left-hand columns are the toxic models and the right-hand the mimicking swallowtails.

7 A 'bird-dropping' spider (*Celaenia excavata*) on a mangosteen tree in Cooper Creek Rainforest, Queensland, Australia.

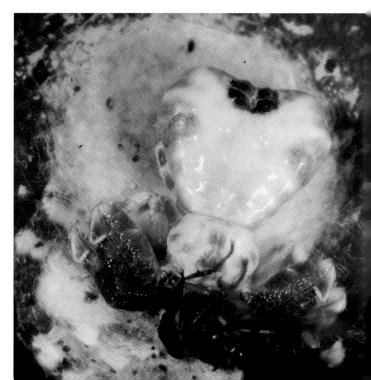

8 The gunboat HMS *Kildangan* in dazzle camouflage, 1918.

9 Abbott Thayer's *Peacock in the Woods* (1907), painted to suggest that the bird is camouflaged against woodland scenery.

10 André Mare's sketch of a 'cubistic' camouflaged gun in the First World War.

11 Pablo Picasso, *Woman with Pears* (1909), a painting typical of his cubist phase, showing disruption of form for artistic purposes.

12 The dramatic eyespot warning display of the mantis *Pseudocreobotra wahlbergii*.

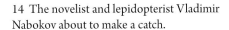

13 In Hugh Cott's photograph disruptive coloration makes these four woodcock chicks almost impossible to make out among the leaf litter.

14 The novelist and lepidopterist Vladimir Nabokov about to make a catch.

15 The lobster moth, a creature credited with mimicking no less than five different species: beech leaves, spider, ant, earwig, or lobster.

16 The best leaf-mimic on the planet – when at rest, *Kallima* holds its fore and hind wings in such a way as to create a perfect leaf.

17 The cruiser HMS *Belfast*, at its current Thames mooring, in the Western Approaches camouflage scheme devised by Peter Scott in 1941.

18 Roland Penrose's 'startle slide' of Lee Miller camouflaged on the lawn.

19 There are two guns in this picture: the visible one is camouflaged in standard 1940 army style, the other camouflaged by the zoologist Hugh Cott.

20 The dummy railhead at Misheifa, Egypt, 1941–2. Everything was made two-thirds size because of shortage of materials.

21 The official New Zealand war artist Peter McIntyre's painting of camouflage dugouts in the Western Desert, 1942.

22 A tank-disguising Sunshield, showing how the casing splits to reveal the tank beneath.

23 The bombardier beetle's orange and black warning stripes advertise a potent, hot, toxic defensive spray. The stream can be directed to any direction by means of a swivelling chamber and the chemicals involved are similar to human rocket propellants.

24 Human warning signs employ the same colours – red, yellow, black and white – that nature uses to advertise dangerous creatures.

25 Miriam Rothschild in her laboratory. Rothschild was the pioneer researcher into the chemistry behind mimicry.

26 A white crab spider on a daisy flower, devouring a bee. To human eyes the spider is hidden but to the bee the flower with the spider is a brighter, more attractive target than a spider-less flower.

27 All of these *Heliconius* butterflies are closely related. The first two rows are *H. numata*, variants on the tiger stripe pattern. Row 3 are *H. erato* and row 4 the *H. melpomene* that mimic them. Despite appearances, *H. melpomene* is more closely related to *H. numata* than to its 'double', *H. erato*. A single genetic switch is thought to be able turn the *melpomene* pattern into *numata*..

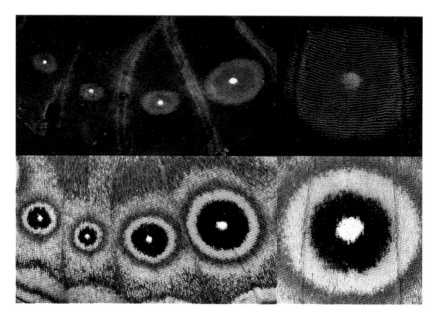

28 How butterfly wing patterns are made – fluorescent proteins signal known gene action that produces eyespots in the butterfly *Precis coenia*

29 *Heliconius hybridisation.* The top two butterflies are the parents: *Heliconius cydno* (yellow band) and *Heliconius melpomene* (red band). The hybrid, *Heliconius heurippa,* is found in the wild and has also been bred in the laboratory.

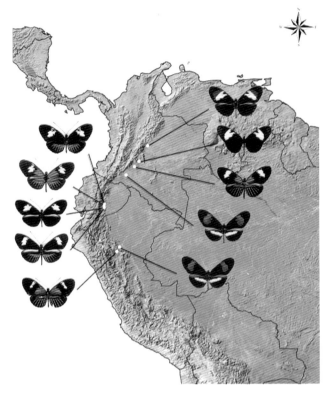

30 The hybrid zone of the Eastern Andes in Colombia with various forms of *Heliconius cydno*, *melpomene* and their hybrids.

31 The Indo-Malaysian mimic octopus, in flounder-mimic mode (left). A flounder in the same territory (right) and the octopus (centre) showing strong disruptive coloration.

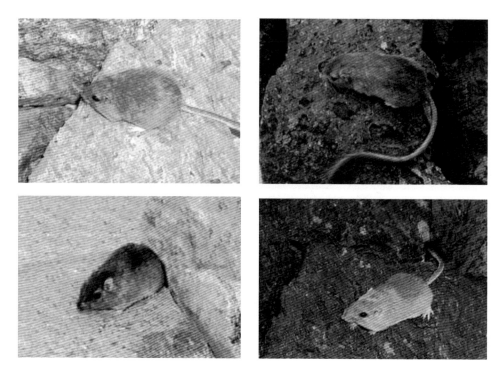

32 The two colour forms of rock pocket mice on matching and contrasting backgrounds.

33 and 34 A toxic coral snake (right) and its harmless mimic, the Sonoran mountain kingsnake (below right). The order of the bands is reversed in the model and mimic. You can tell them apart by the ditty:

Red next to yellow
Kill a fellow.
Red next to black,
Venom lack.

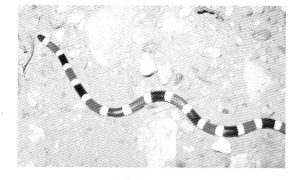

supports. He dismissed a rival scheme at nearby Bassingbourne airfield as 'a childish attempt' – instead of concrete posts, Bassingbourne used 'rabbit wire and wooden pegs stuck into the ground'.

In June 1939, Colonel Sir John Turner from the Air Ministry came to inspect the results. Turner had retired from the Royal Engineers in 1931 and had since then been Director of Works and Buildings at the Air Ministry. He had learned to fly and commanded respect in the Air Force. Norman Wilkinson was inspector of camouflage in his department. Cott was 'somewhat distressed' by the reaction:

> Col Turner (Director of 'works and bricks') came over with Gregory from the Air Ministry on Wednesday to inspect my work, and a rival scheme at Bassingbourne. . . . He told me that it was 'perfect', and that there was no question as to its being the best. But then he regards it as being 'unpracticable' since he believes it would require an artist at every station to direct work on these lines in an emergency, and he regards my work as too expensive, and I gathered that he thinks it's unnecessarily good.

Cott was told by the station commander, who was entirely approving, that 'these air ministry officials are thoroughly jealous, and that they will put up much opposition'. The allegation that his technique would be too expensive particularly incensed Cott because he had been told to spare no expense in this trial run.

But, as the summer of 1939 wore on, the mood improved somewhat. In June, what Cott called 'The Big Six' came to inspect. The party included Sir Kingsley Wood, the Secretary of State for Air, and Sir John Anderson, the Home Secretary and soon to be famous for the bomb shelters bearing his name, which were used during the war. The party was impressed and, as a result, Anderson invited Cott to join the newly formed Camouflage Advisory Committee.

As war approached, Cott was awaiting the final verdict on Mildenhall while correcting proofs of his magnum opus. He had felt fairly certain that RAE Farnborough would want his services after Mildenhall, but on 23 August the bombshell fell. Air Vice Marshal Welsh from the Air Ministry reported:

... although we see no difference in fundamental principle between our two methods and although the specific example you have given us would, we fear, prove too elaborate and costly for general application, nevertheless we are now able to improve our technique within the boundaries of the common principles, thanks to your demonstration, and we intend to make full use of what you have taught us.

In practice, this meant that airfield camouflage would now be an insider job, Colonel Turner's preserve – or would it be Wilkinson's job?

When war broke out on 3 September 1939, Sir John Anderson's committee hadn't met; Cott's book, which he regarded as valuable for the war effort quite apart from its scientific importance, was threatened by cancellation or delay; and much talk about the submarine menace prompted a grumpy note from Cott to Kerr: 'no doubt our friend N.W. [Norman Wilkinson] will be active in this field'.

In mid-October the committee finally met, and Cott came face to face with the opposition. Colonel Wyatt from RAE, Farnborough, he found 'most unhelpful in every way', but he felt he had made some impression on Wilkinson. As a result, he was asked to deliver two lectures at Farnborough. But then in January the axe fell again: Cott was told his lectures weren't needed. He complained to Kerr: 'So it looks as if I shall spend the rest of the war in Cambridge after all.' He was sure that 'N. W. [Norman Wilkinson] is the villain of the piece'. Cott believed that the authorities 'wish to exclude anyone with a real grasp of the subject'.

Having rejected Cott's ideas for the camouflage of real airfields, Colonel Turner turned his attention to designing the decoy airfields. Turner commissioned Gaumont British Studios and Sound City Films at Shepperton to produce dummy Fairey Battles, Hurricanes, and Blenheim and Wellington bombers. After a tricky period with prototypes that were easily damaged in storms, costs that were higher than expected and some opposition from the top brass – six months before war broke out, Fighter Command's Air Chief Marshal Dowding thundered: 'I should therefore most strongly deprecate the slightest diversion of the efforts of my limited Works staff towards the production of dummy aerodromes until real aerodromes have been provided up to my requirements' – hundreds of the dummy planes were produced.

By April 1940 the dummy airfield system was well established. In a memo, Colonel Turner succinctly explained the principle to the Air Commodore with the British Forces in France. There were two kinds of dummy sites: day dummies and night dummies. Night dummies simply used lighting effects only. The game of deception has some intriguing twists. Real aerodromes were camouflaged to make them look like farms; so what should a fake airfield look like? The real giveaways for airfields are the buildings, but resources for the creation of dummy buildings were not available. The answer was that, besides real and dummy air fields, satellite stations were set up with minimal facilities, to disperse aircraft away from the main stations at vulnerable times. The dummies were imitations of the satellite stations, if that is not layering the cake too much.

As well as the dummy night airfields, oil fires known as starfishes were lit in harmless locations after the first wave of bombing raids to encourage subsequent waves to think that these were targets which had been set on fire. Light, of course, is pure illusion. On the night sites, to simulate aircraft landing, a headlamp was set up, running along a wire propelled by a cordite cartridge. Then a second zigzag taxying lamp was provided. Turner recommended putting the lights in crop fields, to save good grassy sites for the day dummies:

> With care and attention the crop can be reaped without disturbing the wiring though the lamps will have to be removed temporarily while the reaping is going on. Later in the season when the ploughing takes place the whole lighting will have to be taken up and relaid if the field is not to remain fallow, but that is looking ahead a bit.

Such concern for the maintenance of the agricultural cycle, even on a dummy airfield in wartime, is peculiarly touching. But, out in the brutal world, France fell two months after this memo was written and for five years lights in the fields of France would come to mean spy drops and the Resistance, and the only airfields would be German ones.

Back in England, the decoys were successful, the night dummies more so than the day. By the end of 1941, the 100 night-dummy landing strips had received more than 350 attacks. During the Battle of Britain, from June to October 1940, they absorbed twice as many attacks as the real airfields. But in December 1940 a map retrieved from a shot-down German plane revealed that most of the day dummy sites had been rumbled by the Germans – only three were marked as genuine. In the last quarter of 1941 even the night sites received only one attack, and by May 1942 the scheme would be abandoned.

Meanwhile Cott was finding the Advisory Committee increasingly frustrating. Although services officers such as Wilkinson attended, they withheld information on the grounds that this committee was essentially a civil defence matter. In March 1940 Wilkinson fobbed the committee off with the excuse that he 'did not know to what extent he could divulge details of camouflage progress'.

Cott was then cheered up by the arrival, on 28 March, of printed copies of his book; but on 5 April he wrote to Anderson tendering his resignation from the committee: 'I am wasting my time.' The committee was a sop, not the real thing. Stymied at home, Cott began to wonder if Palestine or Egypt might offer better pickings. Both Cott and Kerr were outraged at the feeble attempts at camouflage then being perpetrated and campaigned to increase public awareness.

Cott's bid for attention was a long unsigned piece published in *Nature* on 22 June 1940. The article poured scorn on the rash of 'camouflage' that was hastily being rolled out: the familiar lozenges of brown, khaki and grey. Large factory installations such as cooling towers had trees painted on them, as if that would, even remotely, function as concealment when seen from a bomber plane: 'the result of light and shade is such as absolutely to kill this piece of stage scenery at bombing range'. Despite the obvious falsity of the muddy lozenge school of camouflage, Cott realised that 'so consistently has this method been adopted by the authorities that the term "camouflage" has come to mean parti-colouring with drab colours'.

Cott poured especial derision on the camouflaging of London's 5,000-strong bus fleet:

The roofs of many double-decked buses have now been treated with grey or green and brown; but the front, back and sides – the parts most visible in oblique view from the air – are coloured a brilliant red – the very hue the chief attribute of which is its carrying power and conspicuousness.

It is probably true to say that the word camouflage conjures up muddy lozenges in popular consciousness even now. For decades after the war, such patterns could still be seen on factory walls, slowly fading and peeling. But the point of camouflage is to make an object hard to detect against its actual background. Irregular lozenges of earth tones will not necessarily break up the outline of a military vehicle if the paint is shiny, revealing that this is not a natural object. Texture is more important than colour. And such 'natural' disruptive patterns as the conventional green, brown or grey lozenges could be of no use on a bus, which is always seen against a city streetscape. Cott saw all these abominations as '[bringing] down ridicule upon the art of scientific camouflage' (it is interesting that he claims camouflage as both art and science here). Cott attributed these scandals to the 'wrong people' being in charge of camouflage: 'Of sixty-five technical officers [in the Civil Defence Camouflage establishment], all but four are either professional artists or, at the time of recruitment, were students at art schools.' The 'right people', of course, were biologists, and the rightest person of all was H. B. Cott.

The article drew letters pointing out that these hasty examples were the work of amateurs, not of artists, and that artists and scientists other than biologists understood the principles of countershading, indeed that Leonardo da Vinci had anticipated Thayer in this. One letter pointed out, as Cott had not, that air photography, the most important technique in wartime reconnaissance, was entirely in black and white, which nullified colour effects but picked up shadows and changes of textures very readily.

Cott had the last word, as far as *Nature* was concerned. He made the riposte that in August (during the crucial Battle of Britain) such errors were still being perpetrated, pointing to a newspaper photograph of two 12 inch guns, 'each bearing the conventional camouflage pattern which is a peculiarly senseless and ineffective thing, and which entirely fails to break up the surface continuity, to obliterate outline, or to nullify the appearance of solidity due to light and shade'. For Cott, the true response was not more shiny lozenges but countershading.

Cott's thunder did seem to have a galvanising effect, because within days of the *Nature* piece he was commissioned to camouflage two large rail-mounted coastal guns on the East Coast of England. One gun was camouflaged according to standard green-and-brown lozenge practice; the other, by Cott, with countershading *à la* Thayer. In August, the two guns were photographed from the air from different angles and heights and the results were remarkable. In all of the photographs, the Cott gun is invisible except to the most minute scrutiny by someone who knows exactly where to look and what for. The other gun is always highly visible. Cott declared: 'These photographs furnish most convincing proof of the effectiveness of countershading, and are especially valuable in that we have in them a direct comparison between the two methods.' But would the authorities agree this time? Cott awaited their verdict.

The period of ramshackle improvisation and factional infighting among the would-be *camoufleurs* came to an end in August 1940, with the setting up of the Royal Engineers' Camouflage Development and Training Centre at Farnham Castle, Surrey. A strange medley of characters was assembled here, including the artists Julian Trevelyan and Roland Penrose (already installed as a Home Guard camouflage instructor), the magician Jasper Maskelyne, and Hugh Cott himself. At last, the biologists had their man in post.

On 26 October 1940 Cott arrived at Farnham for one month's training before being posted. Life at Farnham was agreeable and Cott began to feel more at home. His expertise was recognised, and he had the chance to lecture to an attentive audience while feeling his way back into Army life. Although *Adaptive Coloration in Animals* was his crowning achievement, he regarded the chance to serve the war effort as far more important.

The authorities were still digesting the results of Cott's gun camouflage. They seemed embarrassed by its success, alleging that it had been photographed 'in a favourable light or angle'. But by December Cott was beginning to sense achievement rather than frustration. Farnham was using his material as the main approach to the subject, and his book was becoming widely read (and even bought) by the officers on the camouflage course. Wyatt saw the point of his countershaded gun at last, and a war artist reported to him from Mildenhall on the success of his camouflage scheme there. As he waited for his posting, he reflected: 'In some respects I should prefer to go off to Egypt or elsewhere overseas for a more varied experience.'

Not everyone at Farnham ate out of Cott's hand. Trevelyan records: 'We laughed at him for his passionate addiction to counter-shading.' The magician Jasper Maskelyne's reaction was even more scathing:

> For six weeks I had to attend lectures where I learned how Arctic rabbits suffer a change of colour when snow falls, and why tigers hang about in tall grass. I had always believed that tigers hung about in tall grass for the same reason that boys hang about at street corners – on the chance of a pick-up.

At Farnham, Trevelyan met up with Penrose again. He was impressed by Penrose's new sense of purpose and by the expertise he had acquired. Penrose had a Quaker background and had grown up with a Thayersque love of nature. He combined the sophisticated artist's approach to deforming the subject with a naturalist's appreciation of animal camouflage. Unlike Maskelyne, he was not inclined to dismiss Cott's examples from nature: in the Penrose archive today, his copy of *Adaptive Coloration in Animals* has a postcard bookmark of *Tanusia picta*, a leaf-mimicking grasshopper, with a sketch by Penrose of its leaf-form on the reverse.

Penrose now occupied a unique position and fulfilled it admirably, bringing some of his surrealist techniques to bear on his various camouflage demonstrations. His most appreciated trick was to drop into his lectures a slide of his lover, Lee Miller, naked on a lawn but for some green camouflage paint and a bit of netting. Penrose gave a straight-faced justification that, if camouflage could hide Lee's charms, it could hide anything. In fact, its purpose in the lecture was more likely an attempt to induce a startle reaction in anyone inclined to doze off.

The Industrial Camouflage Unit might have consisted of mere dilettantes, artistic intellectuals playing at war, but Penrose went on in 1941 to produce an excellent short guide to camouflage, the *Home Guard Camouflage Manual*. Here he correctly stresses that texture is more important than colour and shows how different a close-cropped and a long-grassed field look from the air: 'observation from the air is based on the use of the camera which eliminates colour and accentuates differences in tone'. He notes that the green used in camouflage usually has too much blue in it. 'There is very little blue–green in nature.' Yellowy green with brown is what is required. He had clearly learned from the mistakes of the Industrial Camouflage Unit in its early days, noting how inapposite disruptive patterns are for many large objects. Trucks and pill-boxes merely painted with squiggly patches are not going to disappear against any background likely to be found in war.

The notes of trainees and of those who lectured alongside Cott show that his principles did shine through the entire course. Cott taught that animals are at war every instant of their lives and so are always totally aware of their environment. Human beings, on the other hand, are for large parts of their lives lulled into complacency. As Geoffrey Barkas, who was to become head of camouflage with the Eighth Army in the North African desert, put it:

> What does the average city dweller know about melting into a background and laying an ambush for his enemy? He lives in a world where enormous pains are taken to make everything obvious and self-explanatory. By every device of advertising, lighting, siting, and lay-out the urban man is shown precisely what everything is, where it is, and how to get there. Only a few are ever directly concerned with conceal-ment and ambush – poachers, trappers, dry-fly fishermen, big-game hunters, men on the run – and these do not for long enjoy success unless they are good practitioners of camouflage.

Many soldiers given camouflage netting for a truck apparently believed that it was some kind of invisibility cloak: just throw it over the truck and now you don't see it. Whereas the point is to find a tree, hang the netting so that it forms an awning over the truck which drapes unobtrusively to the ground, then garnish it with scrim and leafy branches. It might seem odd that such things needed to be taught, but they did.

At the very least, the examples from nature were a way of creating vivid images of different kinds of deception, which might otherwise have been an abstract blur. Farnham Lecture No. 5 recognised the following categories:

1 merging (hare, grouse, polar bear etc.)
2 disruption (ringed plover, zebra, moths etc.)

3 disguise (stick insect, leaf insect, sea horse etc.)
4 mis-direction (butterflies, fish)
5 dazzle (grass hoppers, some birds)
6 decoy (angler fish, spiders)
7 smoke screen (cuttlefish)
8 the dummy (flies, ants)
9 false display of strength (toads, lizards, birds).

Cott's *Adaptive Coloration in Animals* must be the only compendious zoology tract ever to be packed in a soldier's kitbag. The book also marks the apotheosis of the descriptive natural history phase of mimicry studies. Although Cott does report experiments on predation to test the efficacy of mimicry and camouflage, the book is essentially a narrative of examples plus theory. After the war, many more experimental techniques were brought to bear on the subject.

But Cott's book is still valuable today for its enormous range, for its passionate exposition of the theories of mimicry and camouflage, and because, in the conflict between artists and biologists, he was both. Cott was a competent illustrator as well as a biologist. Without having Nabokov's precisianism and anti-Darwinism, he brought an artistic sensibility to bear on these phenomena. His text is radiant with the wonder of these adaptations and, fittingly for a book which appeared only two and half months before the disaster of Dunkirk, it conveys a humanistic message.

Cott was very much a follower of Thayer, but without Thayer's excesses. He gives a magisterial summary of the ways nature breaks up the outline, and he may well have been thinking of the Industrial Camouflage Unit when he complained:

> Various recent attempts to camouflage tanks, armoured cars, and the roofs of buildings with paint reveal an almost complete failure by those responsible to grasp the essential factor in the disguise of surface continuity and of contour. Such work must be carried out with courage and confidence, for at close range objects properly treated will appear glaringly conspicuous. But they are not painted for deception at close range, but at ranges at which big gun actions and bombing raids are likely to be attempted.

Cott was unusual in pursuing natural history and camouflage in wartime with equal fervour. In his imagination, the two were interlinked, so that one easily became a metaphor for the other. Richard Dawkins, in *The Blind Watchmaker*, credits Cott with the first application of the phrase 'arms race' to biology; he cites this passage, which more or less paraphrases Belt's argument of 100 years previously:

> The fact is that in the primeval struggle of the jungle, as in the refinement of civilised warfare, we see in progress a great evolutionary armaments race – whose results, for defence, are manifested in such devices as speed, alertness, armour, spinescence, burrowing habits, nocturnal habits, poisonous secretions, nauseous taste, and procryptic, aposematic,* and mimetic coloration ... the perfection of concealing devices has evolved in response to increasing powers of perception, which in many predatory animals, and especially in birds, are of such an order that there is no reason to believe that even the most elaborate cryptic uniforms of tropical insects such as *Tanusia* and *Kallima* have been developed beyond the degree of usefulness.

At times, Cott's sense of wonder at the little narratives concocted by mimetic creatures echoes Nabokov:

> Observations on the habits of some of these creatures read almost like fairy tales – such, for instance, as the angler fish, *Lophius piscatorius*, with his line and lure; or the flowerlike mantis *Idolum diabolicum*, which allures itself to the insects on which it preys; or the American nightjar, *Nyctibius griseus*, which in an upright attitude incubates its single egg on top of the upright tree stump, of which its body forms an apparent continuation; or the masking crab, *Stenorhynchus*, which drapes its body with pieces of weed and other material torn from its surroundings.

Significantly, this passage comes not from *Adaptive Coloration in Animals* but from a talk Cott gave to the School of Military Engineering at

---

* Cott is using the technical terms. Procryptic means camouflaged; aposematic means exhibiting warning coloration.

Chatham in 1938, which was part of his campaign of advocacy on behalf of camouflage. In this passage he was cherry-picking fantastic examples to allure the audience – examples not quite in the naked Lee Miller class, but not without exotic appeal.

The most challenging test for the *camoufleurs* – whether artists or biologists – came in the desert war in North Africa, and many of the Farnham-trained officers wound up there. Hugh Cott arrived in North Africa in January 1941, with the first group of Farnham graduates. They came to the desert with much to learn. Steven Sykes (1914–99), another young artist, wrote how 'units arrived with camouflage nets garnished in greens and browns suited to European landscapes and, as good disciplinarians, they had pegged them out stiffly over the pale sand round their halted vehicles'. Julian Trevelyan observed: 'You cannot hide anything in the desert; all you can hope to do is to disguise it as something else.'

Disguise constitutes one of the three sections of *Adaptive Coloration in Animals*, and many of the natural examples have parallels in human warfare. In deception, illusion is everything. In other words, an object must appear exactly as the deceiver means it to appear – what lies beyond that is irrelevant. Cott put this profoundly in his book:

> The resemblance is literally superficial rather than structural – that is to say, it is generally due to the most ingenious and deceptive disruptive patterns, which give the optical impression of irregular processes and deep interstices – even when painted, as they often are, on the flat canvas of a moth's wing or on the void abdomen of a spider . . . Everywhere we see the same story: the superficial nature of the appearance; and the independent manner of its production.

Cott gives a homely human analogy for this superficiality in the waiter's dickey 'which only resembles a shirt in the limited front area exposed to view.' There is one profession in human life that specialises in creating such waiters' dickeys: film-makers. What appears on screen is all that matters – behind the filmset will be ramshackle braces and struts holding the surface together. It is no surprise that the man who led the desert *camoufleurs* – Geoffrey Barkas (1896–1979) – was a film-maker. Barkas had fought in the Middle East in the First World War and in the inter-war years had made many travel and war documentary films as a producer. Like many others, Barkas entered camouflage work by a mazy route. In a downturn in the film industry in 1937, he had joined Shell-Mex and BP under the legendary patron of artists Jack Beddington. Frustrated in his attempts to join the war effort, he asked Beddington for help. As it happened, Beddington's brother Freddie was setting up the Royal Engineers' Camouflage unit at Farnham.

In November 1941, Barkas established a desert Camouflage Development and Training Camp at Helwan, twenty miles down the Nile from Cairo, with Hugh Cott as chief instructor. Farnham Castle, with its agreeably bohemian collegiate atmosphere, it was not. Julian Trevelyan (surrealist) observed of Hugh Cott (zoologist):

> At the school Cott . . . seems rather lost in this wilderness of sand and little Army huts. In some of them you find petrol cans which you would expect were there as waste-paper baskets, but when you look inside you find snakes, beetles, and lizards, kept there by Cott.

But if a louche artist like Trevelyan remained sniffy about Cott in a traditional, English, two-cultures kind of way, many regular troops had their eyes opened to a new way of perceiving the world. Sergeant Bob Thwaites was a typical NCO who approached the course with a bluff no-flies-on-me attitude:

> Our first acquaintance with our instructor was not encouraging. We had been told he was one of Britain's most eminent naturalists and appeared to have been dragged protesting from a twitcher's hide, bundled into a captain's uniform made by a blind tailor and posted to Maadi.
>
> He was middle-aged, balding and with a military bearing suggesting that he could well have thought Sandhurst to be a seaside resort.

But Thwaites and his group were won over. Cott taught them the preter-natural animal awareness of surroundings and impressed them with observational tricks such as the ability to tell the time of day on a recon-naissance photo which includes a church: many churches are built on an east–west axis, with the altar at the eastern end, so that, effectively, the church is a compass whose shadow tells the time. There were not many churches in the desert but, like Penrose's nude photo, the example produced an impact. More lore that was new to them included the reason why cowboy jackets end in ragged leather strips copied from the American Indians: 'When prone, stalking prey or their enemies, the leather strips would fall to the ground from their shoulders and their arms, thus concealing the shade of their bodies and offering better visual protection.'

Ever true to Thayer's principles, Cott even had an illuminated Thayer box with a countershaded salmon. Disappearing salmon in the desert must have seemed an exotic treat to hot and dusty troops (even if they would rather have been eating one).

Cott also had animal stories to tell. One of his favourite camouflaged creatures was the bittern, a perfect example of one of his key dicta: that camouflage is not just a matter of materials but also of behaviour. The bittern sits in reed-beds with its neck extended vertically. Cott spoke of approaching a motionless bittern among the reeds and gently pulling its beak. The bird responded perfectly when its beak was released by swaying its neck in unison with the reeds. An ideal anecdote for men who might forget that their camouflage netting has to be applied in character with the surroundings, and not just slung over the vehicle.

Cott took them to a military refuse dump and showed them how to retrieve useful camouflage materials. Thwaites again:

> And so on, and so on, while we found ourselves living in a new heady world of make believe . . . By the time the course ended we were a bemused group of men. Some who thought they'd scrounged themselves a cushy break from active service, now looked at the world with different eyes and a new mentality through observing and making instant decisions; using, perhaps for the first time, a form of lateral thinking.

Less enchanted by Helwan (a 'god-forsaken spot') was the most notable English poet of the Second World War: Keith Douglas. Douglas came to the Middle East to fight and, 'without his having made any move himself, he was posted to a camouflage course near Cairo'. At Helwan he found the breakfast 'literally filthy' and filled the complaints book with 'vitriolic but constructive suggestions'. Douglas spent most evenings in Cairo with a French–Egyptian–Jewish family and their five attractive daughters. In February 1942 he wrote that he was now an expert camouflageur, 'though (of course) no one is in the least interested in my qualifications, or in having anything camouflaged'. Douglas's attitude was one of 'I'm a fighting solder, get me out of here!' Eventually he played truant, quitting his post to rejoin his tank regiment in the closing battles of El Alamein. He wrote a searing account of tank warfare in the desert: *Alamein to Zem Zem*.

The Camouflage Unit in the desert was a strange mix of zoologists and artists as well as military personnel. One member of the group, however, deserves further comment: the flamboyant magician Jasper Maskelyne (1902–73), whose contribution to the various projects of disguise and dissembling was either absolutely central (if you believe his account and that of his biographer) or very marginal (if you believe the official records and more recent research). However, he continues to exert a hold on the imagination, as a 2004 Channel 4 film, *Magic at War*, demonstrated (and the persistent rumour of a Hollywood biopic starring Tom Cruise serves to reinforce).

Maskelyne came from a long line of stage magicians and he claimed descent from the famous eighteenth-century Astronomer Royal, Nevil Maskelyne, although this has been disputed. The Maskelynes were inventors as well as magicians; his grandfather John Nevil Maskelyne invented the penny-in-the-slot toilet, which may have disappeared but the phrase 'spend a penny' lives on. It is irrefutable that Maskelyne was employed in the unit and it seems that his primary use was in creating gadgets for escape and forgery on the basis of his own expertise as a magician well versed in counterfeit. Maskelyne's trickbag included tools disguised in a cricket bat, a saw blade inside a comb, and maps on the backs of books and playing cards.

It was in Maskelyne's nature to perpetuate the myth of his own inventive genius, and perhaps he even believed it himself. His ghost-written memoir, *Magic-Top Secret* (1949), makes extravagant claims of cities disappearing, armies re-locating, dummies proliferating (even submarines) – all as a result of his knowledge of the magic arts:

so we got to work again, producing dummy men, dummy steel helmets, dummy guns by the ten thousand, dummy tanks, dummy shell flashes by the million, dummy aircraft; and disguising the real tanks as trucks, disguising Bofors as trailers, making such a colossal hotchpotch of illusion and trickery as has never been accumulated in the world before.

Maskelyne's lurid account was later further embellished by *The War Magician* (1983), written by David Fisher, clearly under the wizard's spell. In the run-up to Alamein, Fisher's account has the army commander Montgomery exhorting Maskelyne: 'I hope you've brought your magic wand with you. We're going to need it.'

Be that as it may, it seems fitting that Maskelyne's precise contribution is shrouded in mystery. Was it real or was it hocus pocus? An official Camouflage Report dated 28 February 1942 shows that, after being head of the small Camouflage Experimental Section in Abbassia, he was 'transferred to welfare', that is, entertaining the troops with conjuring tricks. Who, however, can resist the legend of the *camoufleur*–conjurer who single-handedly won the battle of El Alamein by disguising the troops and their equipment and making them disappear and reappear?

While Cott was teaching lateral thinking in the desert, his mentor Kerr was unhappy. He had become locked into one mode only: perpetual whinge. He regarded Cott's posting to the Middle East as 'a serious scandal' and he wrote to Deputy Prime Minister Atlee to tell him so: 'instead of being retained in this country to train and supervise, [Cott] has been sent to an overseas theatre of war'. And why did the MOD not 'at once take steps to have every gun in service immediately painted as a replica of the Cott gun'? Letters to various military top brass followed, requesting that Cott be brought home. But Kerr was told that Cott could not be released from the valuable work he was doing in the desert.

If Kerr was fuming at home, the troops in the desert needed to find ways of staying sane. Inevitably, this involved humour. Geoffrey Barkas edited a newsletter for the *camoufleurs*, the *Fortnightly Fluer*, in which Cott was gently teased in light verse. A series of verses and cartoons called *Dumb Animals* observed that, while they, the *camoufleurs*, were being urged to 'copy nature' by using strict camouflage, many animals seemed to get by without it:

Observe the owl. Its plumage, see,
Grows lighter UPWARDS gradually,
Thus emphasizing, undismayed,
TELL-TALE effects of light and shade.
   A pity is it not?
. . .
These creatures group with that strange brute
The Contra-Countershaded Newt.
Their origin is in dispute.
Some people think they constitute
   Nature's reply to Cott.

Fun aside, these squibs do make an important point. The lay person sometimes finds it hard to accept the diversity of nature. There are as many life strategies as there are creatures, and the fact that one creature does not use camouflage doesn't invalidate the fact that another one does. The *camoufleurs* had discovered that biology is the science of exceptions.

Besides teaching lateral thinking to NCOs and sceptical artists, Hugh Cott managed to keep his hand in as a zoologist. In October 1941, during a week's leave at Beni Suef on the Nile, he was preparing bird skins when he noticed some hornets busy about the carcasses of discarded birds – kingfishers and palm doves. The hornets seemed to leave the kingfisher carcasses alone and consumed only the dove flesh. This gave him that idea of reversing the usual scenario in which vertebrates, especially birds, prey on insects, which may or may not be unpalatable and protected by warning coloration. Perhaps some birds were unpalatable and signalled the fact with warning coloration? When the war was over this subject became an obsession and Cott never returned to his pre-war mimicry work.*

---

* Cott's papers at the Cambridge Zoological Museum reveal that he did keep notes for a future, revised edition of *Adaptive Coloration in Animals*; but this was never published. He had some amusing human analogies to add, noticing an equivalent of Batesian mimicry in food: margarine being dyed the same colour as butter; 'jellies made more attractive by taking on the colour of the fruit they are almost flavoured with'. Besides the example of the waiter's dickey, which appears in his book, he cites several examples of superficial deceptive appearance in clothing. 'Mimetic Stockings' – painting the illusion of stockings by drawing a seam of eyebrow pencil or mascara down the back of a woman's leg – was a well-known trick in the Second World War; but the most fantastic example is 'Mimetic Pregnancy', gleaned from a book called *The History of Human*

Back on duty, Cott was finding from air reconnaissance photos that, in the glare of the desert, many attempts at camouflaging gun positions with netting actually made the positions more conspicuous. White calico netting made the difference, the point being that the netting needed to be lighter in colour than you would expect.

But, even more importantly, disguise and decoys were the order of the day. In the autumn of 1941 a large operation was mounted which required the creation of a new 70-mile railway line to bring military supplies offloaded from ships to a new railhead for distribution to the forces at Mishiefa. The arrangement was circular, so as to allow for a rapid turnaround, and the area covered was enormous. Steven Sykes recounts being accosted at a meeting by the commander in charge: 'And who the hell are you?' 'Camouflage, Sir.' 'How are you going to hide this lot then?' 'The only thing I can suggest, Sir, is to make a decoy railhead.' Sykes proposed a nine-mile dummy extension from the railhead in a west–north–west direction, ending in a terminus with dummy sidings for discharging tanks and other equipment.

The real railway was being built by the New Zealand Railways Construction Company (the Commonwealth nature of the desert war was very strong: apart from New Zealanders, there was a South African Camouflage Corps, and also many Indian troops). As it happened, there were some surplus rails available; but creating a nine-mile rail extension with tracks, locomotives, wagons, and stores from what was available in the desert was a formidable challenge. These difficulties resulted in the whole railway being on a two-thirds scale (they hoped the Germans wouldn't notice, even when a few real, full-scale vehicles were included), and the last few miles of track were made from beaten-out petrol cans. A cookhouse stove produced smoke for the locomotive. The wagons were made mostly from palm fronds, a ubiquitous building material in the desert.

Stupidity: 'Before the birth of the son of Marie Antoinette, the fashion in pregnancy spread through the Court. The Queen's ladies-in-waiting wore skirts stuffed with cushions to make themselves appear *enceinte*.' Ever attuned to analogies between nature and warfare, Cott noted that the anti-aircraft barrage was actually invented by birds. Some birds such as male eider ducks defend themselves against dive-bombing skuas (that is, skuas, not Stukas) by packing together and throwing up a water barrage. Cott died in 1987; there was a retrospective exhibition of his pen and ink drawings of animals at Cambridge in 1990.

Flares were prepared to create the illusion of damage if bombs were dropped; anti-aircraft guns were emplaced to provide the illusion of defence; and, by mid-November 1941, the *camoufleurs* waited to be raided. Geoffrey Barkas said: 'I think that camouflage men must be among the few otherwise sane beings who yearn to be bombed.' On 22 November the railhead was attacked. Perhaps nine bombs were dropped, but all the eleven flares were ignited at once, in over-eager reaction. On 1 December, the camoufleurs were relieved to come into possession of a captured German map which showed the dummy as the real railhead.

Much later, on 2 April 1942, Julian Trevelyan visited the dummy railhead with Steven Sykes:

> The dummy railhead looks very spectacular in the evening light. No living man is there; but dummy men are grubbing in dummy swill-troughs, and dummy lorries are unloading dummy tanks, while a dummy engine puffs dummy smoke into the eyes of the enemy.

The next day Trevelyan's car broke down and was strafed by three planes. He hears that everything in the area got shot up except the dummy railhead. Later, he heard a story that the Germans had dropped a wooden bomb on it in tribute. Major Stephens, a visiting camouflage officer from India, was more impressed. He quoted from intelligence reports that the dummy railhead

> was most successful in attracting a considerable weight of enemy bombing . . . It is evident that some enemy aircraft were well deceived, as many bombs fell on the dummy position as fell on the real railhead. . . . Captured enemy intelligence maps have revealed that the dummy position was considered to be the real railhead.

The dummy railhead stimulated a flight of fancy in Steven Sykes: a possible disguise whose roots in nature he seemed unaware of. This was a 'two-faced locomotive'. Enemy pilots had learned how to target the vulnerable boilers of locomotives. The idea was to disguise the water-holding tender (which was almost as long as the boiler) as the boiler and vice versa, reverse the engine, and hope that the pilots would hit the wrong end. This ingenious contraption was never tried but had it been implemented, the locomotive should have been named 'Eyespot', because the principle of diverting the attention of predators from a vulnerable to a less vulnerable part is the same as the rationale behind equipping butterflies with eyespots on the wings, where bites will not be fatal.

Another leaf from nature's book was the distressing of the Tobruk distillery. In the desert, fresh water was so vital that protecting the distillery was essential. How could one do it? By pretending that it had already been attacked. After an unsuccessful raid in April 1941, a team set to work producing dummy damage. An illusory hole was created by painting a gaping void in oil and killing the shine with coal dust. Bomb craters were dug around it, debris strewn about and shrapnel holes painted on the walls. Here were *camoufleurs* mimicking a natural ruse that had drawn Nabokov's admiration: 'Consider the imitation of oozing poison by bubblelike macules on a wing (complete with pseudo-refraction) or by glossy yellow knobs on a chrysalis ("Don't eat me – I have already been squashed, sampled and rejected").'

All of these tricks were in a sense vamps-till-ready, dummy runs for the first coordinated use of camouflage as a prime aspect of a major offensive. In early 1942, having fought off Field Marshal Rommel (1891–1944) at Alam Halfa, the 8th Army was rapidly gaining strength for a major offensive. The code-breakers at Bletchley Park in England had cracked the German Enigma codes with their Ultra machine, and this enabled them to sink a high proportion of Rommel's would-be supplies, giving the British forces an advantage in terms of equipment.

Events moved rapidly following Field Marshal Montgomery's (1887–1976) arrival as the 8th Army commander. His great offensive, Operation Lightfoot, to attack Rommel at El Alamein, was scheduled to begin on 23 October 1942. On 14 September, six weeks before the start of the offensive, Colonel Dudley Clarke (1899–1974) visited Montgomery's headquarters at Burg-el-Arab to discuss arrangements for deception. Clarke was Britain's unsung hero of all things deceptive in the Second World War. A theatrical figure who turned his love of masquerade into a dance through the theatres of war, especially in North Africa, he led a charmed life, even surviving being arrested by the Spanish police dressed as a woman. He founded the commandos and was the head of A Force, a shadowy organisation involved in deception of all kinds: Clarke ran agents, spread counter-information and coordinated camouflage deception schemes. It was agreed that A Force should produce a strategic plan for the forthcoming battle. The intention was to strike in the north, in the light of the full moon, on 23 October, while leading the Germans to expect an attack in the south on 6 November.

Geoffrey Barkas and his number two Tony Ayrton were summoned to headquarters on 17 September and given a month to prepare a plan. Clarke had by then left for America, to plan the Allied landings in Tunisia, and the briefing was in the hands of Montgomery's Chief of Staff, Brigadier Freddie de Guingand (1900–79). The first draft of the deception scheme, code-named 'Bertram', was produced on that day. Barkas recalled:

> Ayrton and I trudged through the glaring white sand, seeking a place where we could talk freely without being overheard. . . . Choosing the summit of a dune so that our words would be drowned by the thunder of the rollers, we tried to get the thing in perspective. . . . In two hours, after a feverish session on an ancient and gritty typewriter, Ayrton and I returned to Brigadier de Guingand (Brigadier General Staff, 8th Army) with an appreciation and report.

The brief for Operation Bertram was:

Object:
a) to conceal from the enemy as long as possible our intention to take the offensive
b) when this could no longer be concealed, to mislead him as to the date, and the sector in which our main thrust was to be made.

Because the desert is so open, so pitiless in exposing troop positions, the *camoufleurs* decided to make the most of what they had learnt in order to create many decoys. The strategy required creating a dummy army in the south, just north of an impassable piece of desert known as the Quattar Depression and somehow concealing the real force in the north (Fig. 10.1). Another way of putting it was that they were going to make

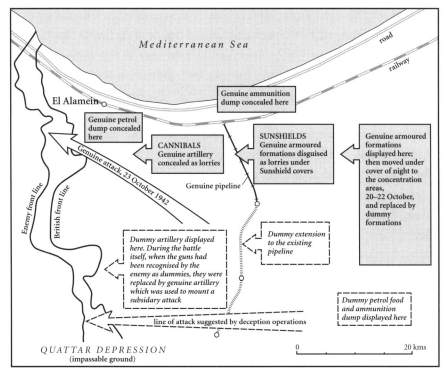

Fig. 10.1   Operation Bertram: the deception plan for the Battle of El Alamein, in the Western Desert of Egypt, commencing 23 October 1942, showing the dummy forces near the impassable Quattar Depression in the south and the disguised forces in the north where the actual attack was made. (Based on a map in NA/WO 201 2024.)

fact follow fiction. The force in the north was going to attack just as Malcolm's army had in *Macbeth* – Birnam Wood would come to Dunsinane.

Here is the point where the equivalent of nature's protective resemblances came into use. Twenty-five pounder guns to be used for the initial assault were ingeniously turned into trucks. Together, the gun and its ammunition trailer easily made a four-wheel truck when they were draped with a fabric body. These were nicknamed 'Cannibals', being lashed together from two pieces of equipment and a fabric cover. Tanks were similarly turned into trucks by means of coverings nicknamed 'Sunshields'.

The Sunshields were the most successful and most exhilaratingly mimetic of the deceptions practised in the desert. They were at first a glimmer in the eye of the Middle East Commander in Chief, General Wavell. On 23 April 1941 he had scribbled a note, a facsimile of which is preserved in the War Office files: 'Is it a wild idea that a tank could be camouflaged to look like a lorry from air by light canvas screens over top? It would be useful; during approach march etc Please have it considered.'

This was passed to the Camouflage Experimental Section and a covering was made up with a wooden frame and canvas. The first Sunshield was in the desert less than two weeks later, on 3 May 1941. There were several problems with imitating a three-ton truck. As Cott often remarked: it isn't enough for something to look like the model, a mimic has to behave like one too. Tanks leave tell-tale tracks compared to a lorry. To overcome this obstacle, a dragline was made from tank tracking; trailed behind the Sunshield, this dragline smoothed the path behind. In the final version, the cover split into two down the middle and the Sunshield could become a tank again in five seconds. Four hundred Sunshields were ordered instantly. And they were to play a critical role in the battle ahead.

Actually the Germans might have been surprised by the number of trucks in the desert, because foodstores were also turned into trucks by being packed tightly into a truck-like shape and covered. As in nature, many distinct species – 25-pounder artillery, Sherman and Crusader tanks, foodstores – all became another species: trucks, trucks, trucks, trucks and trucks. The point of Operation Bertram was to achieve mass deception: an armada of Sunshields and Cannibals was designed to deceive the enemy. It used 722 Sunshields, 360 Cannibals, 500 dummy tanks, 150 dummy guns, 2000 dummy transport vehicles.

The official records of Bertram do not mention Hugh Cott as having any input into the plan, but it is likely that the idea of swapping the Sunshielded tanks and dummies owed something to him. In the 'Disguise' section of his book, he writes: 'Many animals habitually masquerade in garments borrowed from their surroundings; sedentary species using local material, while various wanderers instinctively reclothe themselves from time to time from their more extensive wardrobe.'

This is known as adventitious coloration. A typical example is offered by the caterpillar of the British blotched emerald moth, which, as soon as it emerges from the egg, constructs a disguise of leaf fragments by attaching them to the bristles on its back. Later, after hibernation, it clothes itself in oak bracts. An even more dramatic example, not mentioned by Cott, are the caterpillars of camouflaged looper moths of the genus *Synchlora*, which simply snip petals off with their mandibles and stick them to their back. Adult *Synchlora* are striking green moths, but their caterpillars are even more colourful. Different species use various flowers, but the favourites are yellow flowers such as fleabane or goldenrod. The caterpillars have spines on their back to which they attach the petals with a dab of silk. On a larger scale, many marine organisms drape themselves with seaweed or shells or gravel.

Throughout his book, Cott makes human analogies wherever possible, and here he mentions a police investigation which involved officers who concealed themselves in a tradesman's carrier tricycle to observe a house. He goes on to cite the most legendary ancient use of this principle in war:

> Such tactics offer wide scope in the field of deceptive aggression, whether in peace or war, and as everyone knows, were made famous by the exploits of the Greek heroes who concealed themselves in the gigantic hollow horse at Troy, and thus gained admittance to the city.

In 1943 the Farnham Lecture Notes explicitly made the connection between natural disguise and the Alamein Sunshield deceptions:

> In nature many animals make use of local covering material enabling them to approach their prey unrecognised, as for example various crabs which habitually masquerade in a garment of seaweed borrowed from their surroundings.
>
> In other cases, the prey is actually attracted towards its enemy by a bait or decoy. Certain spiders, for example, closely resemble the dropping of a bird and thus obtain the insects which come to feed at such material; while certain preying mantids of the tropics closely resemble flowers, and thus draw their prey within easy reach.
>
> If a line of tanks moving up into a preparatory zone can pass as some type of ordinary transport and keep their real identity hidden until the very last moment it is obviously a feat of visual deception of the greatest value.

But the true genius of Operation Bertram went beyond the transformation of object into object: these disguised weapons were switched like chess pieces with dummies in a complex game of bluff and double bluff.

In the north, the build-up was concealed by first erecting empty Sunshields where the tanks would eventually take up their posts. On 21 October the tanks were moved up and smuggled under the Sunshields at night. Where the tanks had been, a dummy force was substituted. German reconnaissance next day would have seen nothing new. But a deadly force was now assembled. Alongside the Sunshields were Cannibals for the opening barrage. In the south, a shadow force was established, with dummy tanks, guns, stores and other equipment.

When all was in place, Rommel was not sure which was real – the north or the south force – and he divided his already depleted armies accordingly, in case both were authentic. Mistake number one. Another element of the British strategy was to lay clues to an attack date well after the real one. They built a fake water pipeline from the north to the southern depot. This was timed to be completed well after the attack date, the enemy being encouraged to chart its progress day by day and to calculate the estimated completion date. Rommel seems to have fallen for this – he was ill with tonsilitis and, believing the attack to be some time away, returned to Germany for treatment.

At 21.40 pm on 23 October 1942, the 'cannibals' opened fire. The warlike Montgomery rhapsodised: 'It was a wonderful sight, similar to a Great War 1914/18 attack. It was a still moonlight night, and very quiet. Suddenly the whole front burst into fire; it was beautifully timed and the effect was terrific.'

This disquieting passage reveals Montgomery's ferocity. He believed very much in war by means of deception: outwitting the enemy by strategy and tactics, feinting, playing bluff and double bluff. But in the end he was astonishingly aggressive: 'There will be no tip and run tactics in this battle; it will be a killing match; the German is a good soldier and the only way to beat him is to kill him in battle.'

The Sherman tanks leapt out from under their Sunshields and the battle was on. Although tactical surprise had been achieved, the next phase was not so simple. A minefield had to be negotiated, and a task which was meant to take one day actually took several. But the fact that

Rommel's equipment was much depleted and spread between the northern front and the decoy southern front meant that the Eighth Army prevailed. Rommel didn't move two armoured divisions from the diversionary south to the northern battle zone till 25 October – far too late.

The debate over the contribution made by camouflage to the successful El Alamein battle can never be entirely resolved, but Churchill himself paid tribute to it in the House of Commons:

> By a marvellous system of camouflage, complete tactical surprise was achieved in the desert. The enemy suspected, indeed knew, that an attack was impending, but when and where and how it was coming was hidden from him . . . The 10th Corps, which he [the enemy] had seen from the air exercising fifty miles in the rear, moved silently away in the night, but leaving an exact simulacrum of its tanks where it had been, and proceeded to its points of attack.

Geoffrey Barkas, in charge of the camouflage operation, concluded:

> Though none of us was so foolish as to think that it had been won by conjuring tricks with stick, string and canvas, we could at least feel that we had earned our keep. As an assignment it had been a camouflage man's dream.

CHAPTER 11

DAZZLE (REVISITED) TO D-DAY

*As a boy of twelve I spent a great deal of time studying Thayer's great illus-
trated book on camouflage and was much influenced by it. Later on I became
a keen duck hunter and used a duck punt that was camouflaged in accordance with Thayer's principles of negative shading. Later yet, when serving in
a destroyer in the North Atlantic I was confident that these same principles
applied in the tactics of night attack on convoys by German U Boats and that
the escort vessels could make themselves much less visible by the use of pale
colors and negative shading.*

Sir Peter Scott, Letter to Mrs Mary F.
Boynton, 1950

Abbott Thayer's and John Graham Kerr's idea that the principles
of protective coloration in nature might be applicable to the camouflaging of ships during wartime seemed to have been comprehensively
dismissed by the end of the First World War. The dazzle painting system
devised by Norman Wilkinson had been widely used, but with uncertain
results. After the war, ships reverted to all-over grey. The Second World
War began with no clear policy, and some *ad hoc* colour schemes were
devised by individual ships' captains, but by the end of the war the principles of Thayer and Kerr had been widely applied to warships in the
British and American navies. What had changed? How had this come
about?

What is strange, given the naval authorities' disdain for the naturalists' schemes during the First World War, is how easily the same schemes came to be accepted in the Second. Perhaps the idea of nature's camouflage having anything to teach the military had been too provocative an idea to take in at first, and Thayer and Kerr were the worst kind of advocates, being shrill, prickly, dogmatic and, in many ways, simply unpractical men. Whatever the reasons, the ideas of disruptive coloration, countershading and reduced visibility that were dismissed in the First World War were adopted by both Britain and America in the Second.

Just to complicate matters, the British and American navies interpreted camouflage differently. Having spent the First World War dismissing the possibility of making a ship invisible and investing heavily in Norman Wilkinson's dazzle painting scheme for confusing a submarine's aim, the British became advocates of invisibility in the Second World War, while the Americans plumped for confusion of aim more or less on Wilkinson's principles.

There was no public recognition of any kind of U-turn. The protagonists of the First World War were off the scene. Thayer was long dead. Before the war, Wilkinson had been told that his services were not required in naval camouflage because there would be no dazzle-painted ships in the Second World War: all-over grey would suffice. Kerr was busy working through Cott to get his camouflage principles accepted; but the focus was on the army and air force, not the navy. Insofar as Kerr was still interested in naval camouflage, his aim was to gain official acceptance for his 1914 camouflage proposals.

Any new initiative had to come from a fresh source, and so it did. In Britain, the naturalist Peter Scott (1909–89) succeeded in winning acceptance for Thayer's ideas. There was an intriguing parallel between the two figures involved in naval camouflage. Like Norman Wilkinson in the First World War, Peter Scott was a painter and a seaman who produced camouflage schemes which were widely used in active service. There were similarities and differences. When not at war, Wilkinson was a painter of marine landscapes and boats and a passionate amateur yachtsman. His interest in camouflage was emphatically *not* inspired by nature, and he was absolutely opposed to the idea of 'invisibility' camouflage schemes.

Peter Scott was the son of Captain Scott of Antarctic fame. Peter was two years old when his father died returning from reaching the South Pole in 1912. Scott's mother was a sculptor; he grew up with a passion for nature, especially wildfowl, and for painting it. Scott was something of a gilded youth, being a champion skater, glider pilot and Olympic yachtsman, and he studied natural science at Cambridge but switched in his last year to the history of art and architecture. Already at Cambridge he was selling his paintings of wildfowl. Scott became famous after the Second World War as a naturalist, conservationist, television presenter, and founder of the Slimbridge Wetland Conservation Centre in Gloucestershire. He was the voice and face of popular interest in wildlife. He was knighted in 1973.

Scott's passion for yachting led him to join the Navy in 1939, and he served with distinction as a dashing first lieutenant and, later, commander of destroyers and gunboats, despite suffering, like Darwin, from terrible sea-sickness in his early weeks at sea. Unlike Wilkinson, Scott *had* read Thayer in boyhood and was inspired to put Thayer's theories into practice.

In July 1940 Scott had his own ship, the destroyer HMS *Broke*, camouflaged. The two sides were treated differently. The starboard was plain blue–grey, with white in the shadows; the port side had 'bright pale colours with the object of disrupting the shape *by day* whilst remaining pale enough to fade out by night'. White was used to cancel shadows on the upper parts. This is fascinating, because Scott hoped to achieve a measure of disruption *and* invisibility, both of Thayer's principles. In a later undated document he said: 'Compromise is usually fatal in a camouflage scheme. Invisibility at night must be the *only* objective.' So there was some ambiguity in his thinking, or perhaps he changed his mind.

Like Thayer, Scott believed in the principle of the invisibility of white ships at night. A look at his sketches – including the plan for his own ship – shows that the patterns certainly involve shapes which could be called disruptive, even though they are in pale blues and greens, and not notably dazzling. He recognised that no scheme could produce invisibility under all conditions. He stressed that the aim was for 'the peak of invisibility' to be 'a certain darkness' – by which he meant a typical night, one not illuminated by the full moon.

Scott's designs attained a vogue among other ships' captains, and several vessels were painted unofficially. In May 1941, Naval Officer in Charge, Londonderry, issued a memo to the effect that all ships in the Western Approaches (that is, the North Atlantic), should be camouflaged according to Scott's scheme: 'Camouflage of destroyers as designed by Lieutenant Scott, RNVR of HMS *Broke* has proved very successful. All ships are to camouflage as directed in the following memorandum, at the next available opportunity.' There followed a succinct point-by-point résumé of the principles, including the sensible advice that 'Any one scheme cannot be completely successful in all weather conditions, so one must camouflage for conditions obtaining on the majority of nights in the N. W. Approaches'.

Scott's scheme – which became known as the Western Approaches Scheme – was as popular as Thayer's had not been: there is no doubt that Scott was a good advocate, being charming and well-connected, whereas Thayer was prickly and deeply insecure. The Western Approaches scheme survived even the allegation that it was in some cases too effective, leading to several collisions between British vessels, including Scott's own ship. Despite this, the camouflage remained. Twenty years after Thayer died in despair, Scott had succeeded in getting dozens of boats painted according to something like Thayer's principles.

Scott had worked empirically, from Thayer's principles, but the question of ship camouflage was given a scientific answer in experiments at the naval section of the Leamington Camouflage Centre in November 1941. Photometric tests showed that Thayer and Scott were correct. In overcast conditions, white or very pale colours are less visible than darker colours. The difference is quite dramatic. On a day when the visual range is twenty miles, a white ship can approach six miles closer than a black ship before being seen.

The centre recommended light colours for home waters and darker colours for the Mediterranean. Although the work suggested that colour is relatively insignificant compared to brightness contrast, it did vindicate Scott's idea that some disruptive patterning in pale colours might work in two different modes:

It is possible that, where designs for overcast days are contemplated, colour may be useful in so far as it will be possible to use different colours of the same light tone to make up the pattern. Such a pattern could possibly produce disruption at very close range (when the colours would be visible and produce a colour contrast) and yet merge into a single light tone at greater distances.

This work convinced the Admiralty that Thayer and Kerr had been right all along. In 1943, in a report entitled *The Camouflage of Ships at Sea*, they said of the proposals received in the First World War: 'The soundest of these proposals, whose best points are incorporated in present-day camouflage practice, came from an American artist, Abbott H. Thayer and a British biologist, Professor (now Sir John) Graham Kerr.' Conceding Thayer's point about 'the white ship', as reiterated by Scott and confirmed by Leamington, they said: 'this unorthodox conclusion' resulted from 'the imagination of the artist' (Thayer) and 'the eye of a practiced observer of nature' (Scott). Sadly, there is no evidence that this verdict was ever communicated to Kerr. In 1942 he was still pleading with the Admiralty to reconsider his claim to have priority for dazzle painting. He was concerned that the Admiralty's official line implied that he was guilty of 'deceiving the House and the public'. He seemed to be at the end of his tether.

In the US Navy, in the Second World War, a bewildering range of ship camouflage was used during the course of the war: thirty-three official patterns, plus variants and further experimental types. Curiously, at the beginning of the war, when Japanese radar was not yet functional, low visibility was sought. A graded camouflage system, Measure 12, was the favourite, offering the best compromise of meeting more conditions of light and shade. So the debate concerning low visibility versus confusion of aim, which seemed to have been settled during the First World War, received new chapters in the Second.

The most interesting of the early Second World War US designs, from our point of view, was No. 16 – also known as the Thayer System in recognition of the fact that it employed Thayer's principles. The painter Everett Warner, a naval *camoufleur* in both world wars, commented:

He was one of the earliest advocates of the light or 'white' ship for reduced visibility at night and in overcast weather and when Colonel

Bittinger and I were seeking a name for one of the painting systems using white and pale blue, and authorized for use in Northern waters during the Second World War, we called it the Thayer System.

It was thirty years after the *Titanic* disaster, when Thayer had pointed out that white bergs were inconspicuous at night, and hence so would white ships be.

The US Thayer System was officially recognised as being similar to the British Western Approaches scheme devised by Peter Scott. Western Approaches used three colours (white, light sea blue, light sea green), the Thayer System two (white and Thayer blue). Like Western Approaches, the Thayer System was designed for the North Atlantic, where conditions are generally hazy, but it was claimed to offer course deception in brighter conditions.

In 1943, while Britain was reducing the range of camouflage to a simplified standard pattern, the US Navy strengthened its ship camouflage with three new patterns for different purposes. Pattern 31 was a jungly disruptive scheme intended for concealment against aircraft in coastal locations. In contrast, Measure 32 was an anti-submarine dazzle system.

In 1944, the progress of the war against Japan meant that surface engagements between ships were at a minimum but attacks from the air and submarines were still a problem as they approached Japan. Confusion as to type and course became most important and almost every ship in the Pacific Fleet except battleships was painted in an avowedly dazzle pattern: Measure 32. Much more restrained than Norman Wilkinson' original First World War dazzle pattern, it was intended to show reduced visibility at long distance – reconciling those hard-to-reconcile goals.

Towards the end of the war, in October 1944, the British and American navies conducted joint camouflage trials with respect to the degree to which a ship's camouflage could cause confusion of aim. Their rivalry and possible lack of objectivity in interpreting the trial results are obvious in the official report. Despite the British Navy's outright hostility to the idea during the First World War, the official dogma by 1944 had become the pursuit of invisibility: 'Camouflage for "concealment" is much more effective than is usually realised and achieves much greater results than are measured in trials and laboratory experiments.'

The differences between British and American goals for camouflage can partly be understood in terms of the respective theatres of war. The Americans were fighting the Japanese in the Pacific with its countless islands: they sought inshore camouflage against vegetation. British camouflage was directed towards the Atlantic and Mediterranean theatres: there long-range reduced invisibility was called for.

The Americans had large-pattern disruptive coloration, which would provide inshore camouflage and also a high degree of confusion of aim in combat. In the tests, the US ships were better at achieving this than the British – not surprisingly since the British camouflage scheme did not intend to produce a dazzle effect. Despite this, the British pooh-poohed the US results and suggested that their success in course deception was less than they had discerned.

It might be thought that by now radar would have rendered visual camouflage redundant, but the authorities did not think so. Radar cannot identify whether a ship is friend or foe, or whether it is the biggest prize of an aircraft carrier or something lesser. Once a radar fix is made, visual siting becomes important. The Admiralty stressed that, radar or not, all ships had to be painted somehow: 'There can in fact be no such thing as an uncamouflaged ship.'

The American verdict on the trials was charitable: 'Each type achieves its primary objective – in the case of the Admiralty schemes, low visibility of the ship: in the case of [our] own, confusion of estimates of the ship's range, inclination or identity.' The Admiralty was more surly, stating that the American schemes were too complicated and their implementation would cause 'administrative difficulties'.

Since, after five years of war, the Allies' two major navies disagreed about ship camouflage, it is safe to say that what had been proved was the inexactness of the art, although there had been a degree of consensus in 1941–3.

A fascinating piece of research was conducted in 1944 at the Leamington Camouflage Centre, work that would have gladdened the hearts of Thayer and Kerr. Countershading – the painting of shadowed areas of the ship white and of areas exposed to bright sunlight dark – had been dismissed by the Admiralty in the First World War as impracticable on ships which quickly became dirty at sea. This became standard practice in the British and US camouflage schemes in the Second World War, but in the 1944 experiments the British naval *camoufleurs* suggested that countershading could be improved: the white area needed to extend beyond the actual shadowed area. The aim of countershading is to flatten the image of detail above deck, the shadows created by the bridge, gun turrets and other paraphernalia. It is this that first makes a ship recognisable on the horizon. In tests, the new countershading was far more effective in approaching the ideal of looking like a uniform grey silhouette. But the war was winding down, particularly at sea. The super-countershaded ship was never needed.

Back on land, the war moved towards the decisive event, the invasion of France. The hero of El Alamein, Montgomery, once again had a key role, but he was by now playing second fiddle to the Americans. The success of the deceptions at El Alamein demanded a repetition on a larger scale. The crucial campaign was the D-Day Normandy landings. By now El Alamein-style deceptions had become routine. Plans hatched by the London Controlling Section, the wonderfully deadpan name for the central deception unit, were remarkably similar to Operation Bertram: once again, a large decoy force was assembled with the aim of tying down enemy forces. The attack would be in Normandy and the invasion force was assembled along the south coast of England, between Newhaven in Sussex and Cornwall. Under a plan code-named 'Fortitude', decoy forces were created in Scotland to give the impression of an invasion of Norway and in eastern and southeast England to suggest that the main attack would be in the Pas de Calais. As the indefatigable Colonel Sir John Turner (who had first camouflaged the airfields for the Battle of Britain and whose department was now known simply as Colonel Turner's department) succinctly put it: 'The plan calls for protective decoys between Lands End and the Thames Estuary, and deceptive decoys in Essex and Suffolk.' Decoy sites were provided near the large troop concentrations using the well-tried principles of dummy aircraft (and now also landing craft) and lighting.

The defensive plan for southern England was divided into three sections. Section 1 concerned the area from Cornwall to Newhaven, where the real troop build-up was occurring. Here decoy installations were set up with fake landing craft and lights, including starfish fires which were lit on the decoy sites after bombing raids, as they had been since 1940. Parts of the country with rich associations were put to new uses. Turner wrote:

> Menabilly was sited in a valley close to the country house on which the story and play of *Rebecca** was based. Dams were placed across the valley to flood it, and a number of lights were sited to shine on the water. It was hoped that this might draw off attack from Fowey Harbour which was crammed with craft.

Section 3 extended from the River Colne in Essex to Yarmouth, and the activity there was entirely fake – part of the deceptive Pas de Calais operation. Section 2, between Newhaven and the River Colne, was intermediate, being basically protective. The formations were real but the section's function depended on Section 3 being up and running.

The dummy landing craft, known as Bigbobs and Wetbobs, were the cousins of the Sunshields and Cannibals of the desert. A 'Bigbob' was a dummy landing craft for tank transport, constructed from special steel tubing and canvas. It was made buoyant by the use of floats and had a number of wheels which enabled it to be moved a short distance across land to the water. A 'Wetbob' was a dummy landing craft for assault troops, made of rubber, which took about fifteen minutes to inflate.

Bigbobs and Wetbobs were less successful than their desert cousins. Colonel Turner, whose concerns went way beyond his nominal responsibility for decoy lighting and fires, noted that in Yarmouth the Bigbobs were being erected in full sight of the local population: any German agent would have realised that the whole eastern section was probably a decoy.

But the real failure of the Bigbobs and Wetbobs was their flimsiness. In North Sea winds they were blown about, beached, and could hardly sustain the illusion of an invasion force ready to move. As it happened,

---

* Daphne du Maurier's romantic novel *Rebecca* (1938) spawned plays, films and television serials.

there was little German reconnaissance at this time. Other deceptions, such as signals traffic and rumours planted by agents, were more successful in confusing the Germans as to the intended area of attack.

The preparations with decoys and dummies went on for almost a year before D-Day. On the day itself, 6 June 1944, a new form of deception was attempted. While ship camouflage was somewhat compromised with the advent of radar, electronic technology brought new possibilities for deception. If an attacking force is recognised first by its radar signature, the ability to create false radar patterns could be invaluable. On D-Day itself, the British set out to create a spoof fleet sailing for France.

It had been discovered that Window – metal foil originally developed to be dropped from aircraft to confuse enemy radar – could, if released in a systematic way, create the illusion of a mass of shipping. To achieve this, aircraft had to fly in two boxes of four, dropping increasing bundles of Window as the coast approached and flying in elliptical orbits, the ellipse representing the broad front of the advancing 'fleet'. The orbit would be completed in seven minutes and would represent ships ranging across 16 miles. After each orbit, the next one had to be 0.7 miles further on, to represent the advance of the front of shipping. This exacting task was assigned to 617 Squadron, of Dambusters fame.

Along with the Window drop there were other diversions. A few real ships were included, towing radar-jamming balloons. The whole was a delicate balance, adjusted to give the illusion of a real fleet attempting to hide itself but remaining visible.

There were two operations: Taxable was aimed at the coast between Cap d'Antifer and Fecamp; Glimmer was aimed at Boulogne. In the event, these ambitious deceptions were perhaps too subtle and complicated. The denouement was supposed to be the prelude to a landing in which the balloons would moor offshore; amplified noises of an invasion force would be played very loud and a smokescreen would herald the rush of troops which would never come. But the combined signature of these fake forces was not powerful enough – or the Germans were not vigilant enough – and the deception went unnoticed.

But the larger deception – the phantom army that the Allies had persuaded the Germans was waiting to follow up the Normandy invasion – was triumphantly successful. The Germans became so convinced that the Pas de Calais would be the Allied target that they held to the fiction until long after the actual attack had begun. And the Allies kept up the pretence for a month after D-Day, with false signals traffic and information leaked from agents to the effect that a second large invasion was being prepared. As a result, nineteen powerful enemy divisions which included important panzer reserves stood idle on the day of the invasion, awaiting an assault which never came, when their presence in Normandy might have told heavily against the Allied attack: just as at El Alamein. The German high command kept the bulk of their reserve forces there for almost two months after the Normandy invasion and refused to release troops from defensive duty there, to prevent the Allies from establishing their beachheads.

Some of the protagonists in this story were there on D-Day. Norman Wilkinson painted the invasion as a War Artist; Steven Sykes was there, camouflaging snipers and making sketches; the poet Keith Douglas was there as a soldier and died by mortar fire on the third day of the invasion.

The Normandy campaign was the first in which large masses of troops wore camouflaged uniforms. Mass production of camouflaged uniforms had begun in Germany in the 1930s, taking its inspiration from the great German forests. The most effective camouflaged uniforms, developed for the elite Nazi Waffen SS, were patented, and denied to the ordinary troops. They were based on three forest patterns: plane tree bark, 'palm tree' (actually more like an ash leaf), and oak leaf.

In Britain and America in the Second World War, until D-Day, camouflaged uniforms were also supplied only to elite troops. In Britain the paratroopers were give the Denison smock, a disruptive, leaf-like patterned outfit which remained in service till 1960. In the USA, a horticulturist and the gardening editor of *Better Homes and Gardens*, Norvell Gillespie, developed the first mass-produced US Army uniform. It was called frogskin and was reversible: a summer pattern (with more green) on one side, a winter one (with more brown) on the other. This uniform was not popular with the troops because the double fabric was hot to wear.

All of the camouflaged uniforms of the Second World War shared certain characteristics. They attempted to blend into foliage by means of natural-tinted splodges of around leaf size. Smaller or larger areas of colour were not used. The similarity could cause problems in the theatre of war. In the Normandy campaign, the US troops were hard to distinguish from the camouflaged Waffen SS units. In the end, it was better to be able to recognise the enemy as the enemy than to be concealed from him. Camouflaged uniforms did not emerge from the Second World War with much credit.

At the turning points of the war, at El Alamein in 1942 and at D-Day, deceptive operations had proved crucial to the Allies' victory. The improvisations of the Desert War represented the peak of ingenuity in disguising the equipment of war with mimetic Cannibals and Sunshields to hide the tanks and guns. By D-Day the balance had shifted towards massive deception of intentions, implemented by leaked false intelligence. There is unlikely ever again to be such a meeting point of nature, art and warfare as there was in the Western Desert of North Africa in 1942.

D-Day led to the liberation of Paris on 25 August 1944. On that day Lee Miller, Roland Penrose's wife (she of the naked camouflage demonstration) was in the city with the Allied forces, on a photo assignment for British *Vogue*. She joyfully met up with Picasso again and, after she had her bottom pinched, the two of them settled down to swap stories of the missing war years. Picasso was delighted to hear of Roland's camouflage work, seeing it in terms of zany sketches out of the Marx Brothers. As Roland's son Tony says: 'Picasso would have imagined Roland turning gasholders into elephants, that sort of thing. The idea of camouflage appealed to Picasso's playful metamorphic nature.'

When Picasso pronounced 'God is really only another artist. He invented the giraffe, the elephant, and the cat. He has no real style. He just keeps on trying other things . . .', he was in the line of Bateson and Goldschmidt. As Tony Penrose says: 'They're all the same thing tried out in different ways. Actually, the zoologists have gone to the trouble of proving this true because the elephant's skeleton has the same number of components as the mouse's.' And not just the skeleton – it goes much deeper than that. It was peace-time again: time to return to biology.

CHAPTER 12

FROM BUTTERFLIES TO BABIES AND BACK

As the laws of nature must be the same for all beings, the conclusions furnished by this group of insects must be applicable to the whole organic world; therefore, the study of butterflies – creatures selected as the types of airiness and frivolity – instead of being despised, will some day be valued as one of the most important branches of Biological science.

Henry Walter Bates, *Naturalist on the River Amazons* (1863)

In the October 1952 issue of the *Bulletin of the Amateur Entomologists' Society*, a small ad appeared which had dramatic but benign medical consequences about fifteen years later. It was also profoundly important for the study of mimicry in butterflies. You can still read the ad today in library copies of the magazine: the small ads are printed on crudely typewritten sheets bound into an otherwise standard printed magazine. The ad read:

Dr P. M. Sheppard (291), Genetics Laboratory, Dept. of Zoology, University Museum, Oxford, wants living eggs, larvae or pupae of *Papilio machaon* (Swallowtail) from the Continent for genetic research. Will buy, or exchange for living British *P. machaon* or South African *P. demodicus.*

Since 1946, Philip Sheppard (1921–76) had been the rising star of the Oxford School of Ecological Genetics under its eccentric leader,

E. B. Ford. Ford was a geneticist who, as far as Britain was concerned, knew more about butterflies and moths than anyone else. The two popular books he wrote for the Collins New Naturalist series – *Butterflies* (1945) and *Moths* (1955) – are still read today.

In the 1950s and '60s, Ford's department was the leader in mimicry research, and very many researchers passed through it on the way to setting up departments across the world. What was special in Ford's work was the combination of ecology and genetics. R. A. Fisher was Ford's mentor and he had stressed the importance of mimicry as a simplified version of the problem of the whole of biology: the wing patterns, being two-dimensional, are likely to be simpler in their genetics than three-dimensional body organs, and for butterflies life consists mostly of evading predators and mating (many butterflies do not feed at all as adults: mating is their sole purpose). Ford never lost sight of this focused approach.

Like all schools in science, Ford's department had to fight its corner: to justify its approach, to gain funding, and to head off criticism from rival factions. We have already seen many personal feuds in this story and mention of E. B. Ford cannot duck the issue of yet more to come. The initials E. B. stand for Edmund Briscoe, but this most formal of men was called 'Henry' by his friends (Miriam Rothschild even called him Henry in her paper for his *Festschrift*, 'Speculations about Mimicry with Henry Ford'). He had many acolytes and friendly collaborators and was a generous mentor, but it is fair to say that anyone who was not a member of Ford's school, or at least an upholder of his broad tendency in evolutionary biology, was inclined to get the sharp end of his tongue. One of his adversaries, J. B. S. Haldane* (they had a grudging respect for one another, although Haldane was a Marxist and Ford a High Tory), commented in a review of Ford's *Ecological Genetics* that he seemed to have picked a quarrel with most of the important figures in biology. He concluded that Ford's defence of a limited territory against all comers 'will perhaps increase the number of readers who dismiss Ford as a

---

* J. B. S. Haldane (1892–1964) was a controversial figure in genetics. A staunch Communist for most of his life, he supported Lysenko until 1950. He risked his life in conducting physiological experiments on himself.

"heresiarch and pseudopontiff"* (to borrow Rolfe's phrase). I am not one of them. I think that Ford is generally right in his more controversial conclusions.'

Ford was a mandarin, a dapper, owlishly bespectacled little man with camp mannerisms, who adored the etiquette of High Oxford. He was a Fellow of All Souls and fought against the admission of women even to lunch. He circulated Fellows with the following:

> I very much hope that you will be able to attend the college meeting on 11 March and that you will give favourable consideration to voting against the Motion, to come up then, of having women to the college lunch. The presence of women would completely alter the unique and valuable structure of All Souls . . .

But Ford was also a serious evolutionary geneticist (albeit with some-what reactionary tendencies). The whole thrust of his work was to show evolution in action by observing the fitness for survival in creatures bearing alternative genes for certain traits. These alternative genes are known as polymorphisms. Ford was the expert on polymorphisms, so let's be clear about this topic. They are not to be confused with muta-tions. A mutation is a one-off chance event, which might, if it is useful, spread throughout the population. A polymorphism is a variant gene that exists to a certain extent all the time in the population. For example, in a sense, men and women constitute a polymorphism. There is no standard human being: 50 per cent come as men and 50 per cent as women. A polymorphism was succinctly defined by Ford as 'the occurrence together in the same habitat of two or more discontinuous forms of a species in such proportions that the rarest of them cannot be maintained merely by recurrent mutation'. Which is obviously true both of men and women and of swallowtail butterflies.

---

* This is a wickedly accurate jibe from Haldane. In antiquity, a heresiarch was the leader of a sect. The phrase comes from Frederick Rolfe's (Baron Corvo's) book *Hadrian the Seventh: The Triumph and Tragedy of a Spoiled Priest*. Ford, with his veiled private life and love of the highest form of Oxford flummery, was indeed a 'spoiled priest'. Ford must have squirmed to have been as accurately pinned down and formulated as one of his butterflies.

In 1936 Ford had published a paper in which he drew attention to the swallowtail *Papilio dardanus* – the butterfly Wallace, Trimen and Poulton had marvelled at – as a model organism for the study of the evolution of polymorphic mimicry. No cross-breeding experiments had been conducted outside Africa at the time, but Ford surveyed what information there was in an attempt to understand the dominance relationships of the various female forms which mimic other, toxic, butterflies. The paper was a call to arms. In 1952 Philip Sheppard took up the call.

The advertisement in the *Amateur Lepidopterist's Bulletin* was answered by Dr Cyril Clarke (1907–2000), a consultant physician at the David Lewis Northern Hospital in Liverpool and a keen amateur lepidopterist. He had been fascinated by swallowtails since childhood, when he first saw them on the Norfolk Broads.

After six years in the Navy in Australia, Clarke celebrated the post-war return to normality by beginning to breed, at first purely as a hobby, the butterflies he had so loved as a boy. In 'captivity' they generally refuse to mate naturally, so Clarke perfected a technique of hand-mating them: female in the left hand, male in the right, the male's claspers opened by a fingernail, then hey presto, they would get down to it. I say 'Clarke perfected' because that's how it appears in the published paper; but it seems that Clarke's wife Feo was the real expert in hand-mating.

Clarke began to correspond with other butterfly breeders worldwide. In September 1952 he was sent some black swallowtails by a butterfly enthusiast in Georgia, USA. In his own words, 'in an idle moment one Sunday afternoon I hand-mated her to a male of the yellow British species'. The mating was successful and the offspring were all black, showing that the black American form was dominant to the yellow British one.

When Clarke back-crossed the hybrid to the yellow British parent, the result was that half of the butterflies turned out yellow and half black. This was the classic Mendelian pattern. The offspring of the first cross are black because they receive one black gene and one yellow gene, and black is dominant to yellow. But the hybrids do have one yellow gene; so, when they are back-crossed to the yellow parent, half the offspring get two yellow genes, and hence the yellow form reappears in the next generation.

This was elementary, but the point was not to demonstrate Mendel's principles once again: what the crossing proved was that, in these two very different species of *Papilio*, the colour patterns seemed to be controlled by a single gene, which existed in two forms – a polymorphism. In 1953 the experiment produced Clarke's first publication on butterflies: 'A hybrid swallowtail'. But meanwhile Clarke had seen Sheppard's advertisement. They began to correspond and, in his obituary notice of Sheppard, Clarke recalls that 'we met for the first time to discuss the matter in the bar of the Mitre Hotel, Philip wearing his ancient R.A.F. greatcoat'.

During the war, Sheppard had been a navigator on the first 1,000-bomber raid on Cologne. His plane was shot down over the North Sea on his sixteenth mission, on 27 July 1942. Sheppard spent the next three years in prison camps, where he took a correspondence course in horticulture in order to stay sane and participated in escape attempts. He was in the famous Stalag Luft 3 camp and assisted in the wooden horse escape – as neat a piece of decoy work as ever took place in the human realm. The wooden horse was a vaulting horse which the crazy English prisoners took out to the same spot every day for exercise. Stalag Luft had been designed to foil escape attempts. The huts were built on stilts to reveal any tunnelling activity and the soil was a grey dust with sand beneath, which meant that disposing of it was exceptionally difficult. The men tunnelled beneath the horse every day. Slats from the huts were used to line the tunnel and to dispose of the soil the prisoners sewed a pocket with a hole into their trousers: as they walked about, the sandy soil trickled out in a fine stream, dispersing against the topsoil without creating a tell-tale coloured pile. Sheppard said after the war that he 'wasn't tough enough to vault six hours a day on prison fare so he helped to carry earth from the tunnelling'.

Clarke and Sheppard began to collaborate on research, for the first few years remote from each other, in Liverpool and Oxford; but in 1956 Sheppard came to Liverpool as Senior Lecturer in Genetics in the Department of Zoology and their work on swallowtails gathered pace. Clarke realised that, besides butterfly genetics, Sheppard's expertise could be brought to bear on medical problems. But in what form? Clarke recalled: 'One day, motoring together to the Norfolk Broads, I asked him for his views. "Blood groups," he said, and blood groups it was.' Later, reviewing their work, he said: 'We could not help noticing certain striking parallels between the inheritance of their wing patterns and inheritance of blood types in man.'

Polymorphism is not restricted to butterflies. In 1900, the Austrian physician Karl Landsteiner had discovered the now familiar human blood groups, A, B and O. These are polymorphisms. There is no standard human blood group type: everyone has one type or another. Both R. A. Fisher and E. B. Ford had wondered if butterfly polymorphism might contribute to a better understanding of the blood groups.

The second similarity between *Papilio* butterflies and blood groups is that in the butterflies something interferes with the expression of the mimicry gene in the *male* – it is switched off. The medical problem that Clarke wanted to solve was that of Rhesus babies. This involves additional blood group factors discovered in 1937, known as the Rhesus system, which only cause problems in *women*. Both *Papilio* butterflies and Rhesus babies involved questions of sex-linked inheritance.

The Rhesus problem occurs when blood from the foetus of a Rhesus-positive baby (Rh-positive means possessing the Rhesus factor; Rh-negative signifies its absence) gets into the bloodstream of a Rhesus-negative mother. When this happens, the mother will start to produce antibodies against the Rh-positive blood cells from the foetus, and these will remain in her blood. In a first pregnancy this poses no problem, but at the second pregnancy the mother has now manufactured antibodies against the Rh-positive factor that entered her bloodstream from her first baby. Whereas foetal blood only enters the mother in rare cases of leakage, most commonly at childbirth, maternal blood is passed into the foetus as a matter of course. So, in a mother sensitised to Rhesus incompatibility in her first pregnancy, her Rhesus antibodies will knock out her baby's red blood cells. Before the Second World War, such babies usually died soon after birth. During the war, techniques were developed to give transfusions to Rhesus babies and this could often save the baby's life. But something better was needed.

Clarke and Sheppard began to apply their knowledge of polymorphic genetics to the problem. As Clarke tells the story, when he was wrestling with the question of how to immunise mother and baby against this scourge, his wife woke him after a dream, shouting 'give them anti-Rh'. Anti-Rh is the antibody generated when foetal Rh-negative blood meets the mother's Rh-positive blood. Clarke recalled: 'In a huff I replied, "It is anti-Rh we are trying to prevent them from making." ' But then he realised that giving the mother antibodies in this way, to remove incompatible Rh-positive cells before her own antibody machinery went into production, might be a viable strategy. The possibility of foetal blood leaking into the mother's bloodstream is only a problem at delivery, so the antibody only needs to be administered at that time.

Tests began on giving the mother the anti-Rh factor, to disable her Rh-positive cells at the sensitive time. On 4 March 1964 the first treatment was reported in *The Times*, and Clarke's team began a clinical trial. The circumstances of the first treatment were slightly odd. An American woman whose husband was a doctor turned up at Barnet Hospital, in North London, requesting the treatment. As *The Times* put it: 'The arrival of the woman in Britain with the serum came as a surprise to the Barnet hospital, who asked advice from Liverpool. She is understood to be related to an American worker in the same field.' The article made sure it concluded by acknowledging the pioneering British work: 'The idea of the new method of protection originated from the Department of Medicine at Liverpool University. The serum given is an improved version of the one developed by the department.'

A tangible endorsement of the power of the idea that grew from Cyril Clarke's innocent butterfly breedings of the early 1950s had come on 7 November 1963, when Liverpool University received a grant of £350,000 from the Nuffield Foundation in order to establish a fully fledged unit: the Nuffield Unit of Medical Genetics. E. B. Ford had been instrumental in securing this grant.

In 1965–6 the results of the clinical trial showed that in 78 treated cases none had the Rhesus antibody six months later; 19 of 78 controls did. In the following year's report a few failures were noted, but Clarke's team felt sufficiently confident to abandon controls and give the treatment to all mothers who requested it. By 1970, this had become a routine technique. The results can be seen in the UK infant mortality figures: in 1950, infant mortality from Rhesus disease was 1.6 babies per thousand; in 1970, it was still 1.2 per thousand, but by the early '80s it had fallen to 0.1 per thousand.

Those airy, frivolous butterflies had made a contribution to human well-being. Here, at least, was one indisputably good deed in a naughty world. When I spoke to Michael Majerus, Professor of Evolution at Cambridge University until his untimely death in January 2009, he remembered the impact that Clarke and Sheppard's work had on him at an early age:

> When I discovered that story, in 1964, I was 10 years old, which was when my parents mistakenly gave me an easy butterfly book which turned out to be a genetics book (Ford) and then 6 months later I'd saved up pocket money so I could get his *Moths*. And when grandma said: why do you collect butterflies, these old eggshells you found around the place? I had no answer. But when I discovered that the Liverpool Rhesus jab came from these swallowtail butterflies I had something to say to grandma.

Sheppard continued his work at the university and was also employed by the Nuffield Unit, with Clarke as director. Clarke now had to add planning a new building to his other skills as a physician and swallowtail expert. The new building was much delayed, but on 1 March 1967 Clarke wrote to Ford: 'The crane of the Nuffield Unit building has been taken down and we are nearly in.' The building (now demolished) was a gloriously 1960s' edifice, on stilts, and punctuated by a block of cubes two thirds of the way down one side.

That such a large, concrete (in both senses) project should have emerged from that obscure little advertisement is a tribute also to the often maligned Oxford School of Ecological Genetics – both for educating one half of the Clarke–Sheppard partnership and for Ford's lobbying for the Nuffield Grant.

The apotheosis of the Ford–Clarke partnership was the opening of the new building on 26 May 1967. An amusing correspondence between Clarke and Ford reveals Ford's excessive punctilio. He insisted on giving a cocktail party as part of the ceremony, although he was guest of honour, and he was much concerned about dress code: 'There is, of course, the general proposition that one never wears the robes of one university in another and consequently I would normally wear no robes, even though everyone at Liverpool was doing so.' But, now that he was an Honorary Doctor of the University of Liverpool, 'would it be possible for the University to lend me the dress robe and academic cap?'

The event was a huge success (Ford to Clarke: 'Everyone is saying there has never been an opening like it') and Ford stayed with the Clarkes. He was shown the breeding butterfly collection and wrote to Clarke's wife, Feo: 'And thank you for the wonderful opportunity of seeing the butterflies yesterday morning. I suppose I shall get them clear in my mind some day, but that time is not yet arrived.' That Ford should have found the swallowtails confusing is reassuring: Clarke's and Sheppard's papers are vastly complicated, and Ford had prepared the way with his paper on the subject in the '30s!

To understand what Clarke and Sheppard were trying to do with *Papilio*, we need to recap. Darwin had embraced Bates's discovery of mimicry as a prime example of natural selection, but he ruminated on the mechanism: in the sixth edition of *On the Origin of Species* in 1872 he speculated:

> It is necessary to suppose in some cases that ancient members belonging to several distinct groups, before they had diverged to their present extent, accidentally resembled a member of another and protected group in a sufficient degree to afford some slight protection; this having given the basis for the subsequent acquisition of the most perfect resemblance.

This was a rather desperate attempt to reconcile the observable fact of mimicry with his theory. He was suggesting that, before the butterflies became so different, for example the cabbage white family and the Heliconidae observed by Bates, they must have had some similarity that enabled natural selection to get to work, eventually producing a convergence in mimicry.

Over the next century, controversy raged: could mimicry have been perfected by a series of small cumulative variations or had there been a large discontinuous (hopeful monster) leap? Punnett in 1915 already suspected that a single genetic locus controlled the mimicry in some cases. He decided that this must have evolved in a single step – as did Goldschmidt in the '30s and '40s. But the dominant school in biology, the proponents of the modern synthesis, also known as Neo-Darwinists, insisted that all evolutionary change came about by way of a series of incremental small mutations. By 1953 Ford had realised that the facts were starkly contradictory. Mimetic polymorphism involves many individual changes of shape and colour on the wings in order faithfully to mimic the models. Yet the entire pattern in polymorphism is controlled by a single genetic switch mechanism. Such examples of polymorphism must have arisen by a single mutation at the locus in question. How would a sudden mutation produce all of the changes required? This is rather reminiscent of the hoary old question: how long would it take a tribe of monkeys to type the whole of Shakespeare – or even *Hamlet*, come to that? The answer is: never. The odds for the butterfly are slightly better but, as Ford said, for the perfect mimetic resemblance to be produced by a single mutation is hard to accept, but take the many cases of polymorphic mimicry and the scenario is, in Ford's words of 1953, 'beyond belief'.

Nevertheless polymorphic mimicry happened: it is our understanding of the genetic mechanism that is incomplete. To solve the problem, Ford advocated a two-step solution, in which a reasonably lucky first strike (well short of a hopeful monster) was gradually refined by natural selection. The mimicry process is like a whirlpool which allows most species to continue serenely and unchanged beyond its outer whorl; but, if a freak resemblance brings a butterfly into the orbit of the mimicry maelstrom, natural selection makes sure the resemblance is steadily perfected. In other words, the over-perfect resemblance that Nabokov regarded as being beyond the power of natural selection is only what you would expect once the process of mimicry had begun.

Also in 1953, Ford made another contribution by pointing out that at least one of Goldschmidt's tenets was certainly wrong. Goldschmidt took up Punnett's idea that patterns could easily be copied from model to mimic because all butterflies had a limited repertoire of possible patterns. Goldschmidt took this further in asserting that model and mimic must use *precisely* the same genetic pathways (obviously, this created a huge gap between mimicry and special resemblance: Goldschmidt did not claim that the genetic mechanisms of a *Kallima* butterfly and a plant leaf were identical). Goldschmidt was probably right in some cases, but Ford referred to the conclusive evidence that the reds of some poisonous swallowtails and the butterflies which mimicked them were chemically distinct pigments; and similarly with some yellows and whites. This is a confirmation of Cott's superficiality principle, according to which the appearance is everything and the means to attain that appearance incidental.

By 1954, Clarke and Sheppard were sufficiently confident to embark on a mammoth cross-breeding programme in swallowtails in order to decide the issue of large discontinuous leaps versus small cumulative variations. Ford suggested that there was a test. Where a model shows variation across a range of territories – subspecies or varieties – the mimic follows it, imitating the changes faithfully. Goldschmidt had rashly and brashly predicted that such changes would be produced by a single polymorphic gene, just as the major mimetic patterns were in the swallowtails. Ford thought this most unlikely and predicted that, when crossing experiments were done, the patterns wouldn't always come out cleanly as one type or another, as they did in swallowtails. Instead, a host of intermediate forms would be produced.

Sheppard and Clarke's grand conclusion, after cross-breeding count-less thousands of butterflies from many parts of Africa and Asia, was that the major mimetic forms of female *Papilio dardanus* – totally different patterns copying tailless toxic species from other genera – were controlled by a single gene: 'As far as colour and pattern are concerned, only one locus apparently is responsible, the presence or absence of tails on the hindwing being determined by the other.' So this was an extreme case of polymorphism. The gene locus involved seemed to exist in about eleven different forms. Each of the eleven had a place in a dominance hierarchy – those high in dominance were thought to be more recently evolved, and the least dominant form was thought to be the first. The tailed male pattern, which is the same for swallowtails of a particular species every-where, was thought to be the ancestral pattern from which the tailless mimetic forms emerged.

How could this be? It is as if a single switch in a house could not only turn on different shades of light in each room on different occasions, but also swap the curtains round, repaint the front and hang a new name plate on the door. At least the single gene for tails only had to switch one thing on or off! How the two changes – for mimetic patterns and for the presence or absence of tails – were made posed intriguing puzzles. Clarke and Sheppard worked well before the era of molecular biology, but, since Watson and Crick's discovery of the DNA structure in 1953, it was believed that each gene was responsible for making a single protein. So how could a change in one single gene – which could only change a single protein – produce all these powerful changes in pattern? Clarke and Sheppard's gene seemed to be some kind of master switch. They called it a 'supergene', and the enigmatic properties of this 'supergene' go to the heart of the mystery of mimicry.

A supergene looked suspiciously like one of Goldschmidt's hopeful monsters, but Clarke and Sheppard asserted that 'This mode of genetic determination does not necessitate that each pattern arose fully perfected as the result of a single mutation, but only that the various allelomorphs* act as "switch mechanisms".' The actual patterns controlled by the switch mechanisms would then have been progressively improved by modifier genes, acting under natural selection on what I have called the whirlpool principle. Clarke and Sheppard did not believe, as Punnett and Goldschmidt had, that the supergene evolved instantly, as a one-off mutation: other genes had become linked to create it.

What of Ford's test? Clarke and Sheppard found that, although the major forms of the swallowtail were controlled by one powerful switch gene, when some of the local races were crossed, a host of intermediates was found: here several genes were involved. They concluded that Ford was right and Goldschmidt wrong.

To bolster their case, they found that other genes were needed to keep the mimicry true. The best evidence for this came from Madagascar, where, as Hewitson had pointed out way back in the 1860s, *Papilio dardanus* exists as a single form, males and females alike – with tails. In 1959, Clarke and Sheppard asked a colleague who was travelling to Madagascar to bring back some eggs and larvae of the local form. These were crossbred with the South African varieties.

The specimens from these experiments, with notes relating them to the published papers, resided for some years – while the new Darwin Centre was being built at the Natural History Museum in South Kensington – in a cool basement of a warehouse in Wandsworth.

---

* An allelomorph is one of the patterns controlled by a gene that displays polymorphism – two or more forms of the same gene.

The butterflies were still in Wandsworth when I visited. Dick Vane Wright, former curator of Lepidoptera, took me down, through a room full of grinning mammals, to see the swallowtail mimics and their models. As Sir Cyril Clarke points out in his notes, this is not a pristine display of perfect specimens, for the simple reason that these were breeding butterflies and, after laying, the females quickly become shabby.

The shabbiness of *some* of the specimens, though, had a greater significance. The offspring of these Madagascan/South African matings are very suggestive. In the first generation, the mimetic pattern of the South African parent comes through, but in a smudgy way, like a poor photocopy. But the tails remain and the whole wing is littered with a scurf of random markings. So, although the principal mimicry gene works up to a point in distant Madagascar, where there has never been any mimicry in this species, some refinements present in the South African race have gone missing in Madagascar. After two generations of Madagascan/South African crosses, the mimicry was even poorer. The manner in which mimicry breaks down suggested how it might have come about in the first place. It showed that a single gene controlled most of the pattern, but that, in populations where there were models to copy, other modifier genes helped to keep the mimicry true. Without the models, the patterns soon drifted out of focus. In Madagascar, there was no genetic whirlpool to drag the patterns into perfection at the vortex.

But in 1975 a husband-and-wife team of evolutionary biologists, Brian and Deborah Charlesworth (now both Fellows of the Royal Society), threw a spanner in the works by showing that Ford, Clarke and Sheppard were wrong: that the separate modifier genes for perfecting the mimicry could not have become linked and incorporated into the supergene, as they had supposed. The Charlesworths showed theoretically that the multiple loci involved in a supergene must have been linked *even before the pattern-making mutations occurred.* This was a terrible blow to the only plausible theory of the evolution of polymorphic mimicry on the table. If perfect mimicry could not evolve in a single hopeful-monster mutation and the genes which perfected the mimicry could not become linked to create the single switch gene, there was an impasse: how on earth had those fabulously mimetic swallowtails evolved?

And *Papilio* bore out the Charlesworths' theory. Remember that there are two genes controlling the swallowtail patterns: one for most of the wing pattern and one for the tails. Given that the lack of tails was necessary for mimicry, the two ought to have become linked. But the Charlesworths showed that this linking could not happen. Depending on whether there are non-mimetic as well as mimetic forms in any one place, all the female *Papiliones* should have tails, or none should have them. In other words, if mimetic and non-mimetic *Papiliones* fly together, the mimetic *Papiliones* would not be able to lose their tails. This is what we find. In sub-Saharan Africa, all the *Papiliones* are mimetic and no female has a tail. In Ethiopia, 60–80 per cent are non-mimetic and *all* of the *Papiliones* have tails.

The paradox uncovered by the rigorous experiments of Clarke and Sheppard and by the equally rigorous theoretical work undertaken by the Charlesworths could not be finally explained until molecular biology came of age: until, that is, the genes could be investigated on a fine scale, down to the level of the chemical bases of DNA. But in 1975 the problem seemed maddeningly beyond comprehension. Why would genes that would eventually come to create the pattern of a totally different butterfly be lurking, ready and waiting, linked even, long before the mimicry began? It sounded like a disaster for evolutionary theory. John Turner, another graduate of the Ford School, faced up to this problem:

> Many people will find it improbable, if not smacking of special creation, that the loci should just happen to be appropriately linked. It is indeed improbable, and that is why there are so few spectacularly polymorphic mimics among butterflies: and only a few species happen to have clusters of loci controlling wing pattern functions. These are the ones that attract our attention by becoming polymorphic: in a way we are performing a biased experiment.

Turner went on to propose the idea that only those butterflies which had a pre-adapted set of linked genes capable of producing polymorphism would become mimics. This would explain Punnett's 1915 puzzle as to why many butterflies are not mimics even when it would be an advantage for them to be so: they don't have a suite of pre-adapted linked genes ready to press into service. This was the position before the modern era of evolutionary developmental biology, which was getting under way just as Turner wrote in 1984; but it would be more than twenty years before the puzzle would be solved.

Following the *Papilio* epic, Philip Sheppard turned his attention, working with John Turner, to the even greater task of trying to unmask the genes behind *Heliconius* mimicry. Sheppard was unable to see this through. In 1976 he developed leukaemia, and the mammoth *Heliconius* paper was unfinished at his death in 1977. 'Genetics and the evolution of muellerian mimicry in *Heliconius* butterflies', by Sheppard, Turner and co-workers ('almost a book', in Turner's characterisation) was finally published in 1985, three years after submission. This work, with its detailed analysis of the genetic loci involved in *Heliconius* colour patterns, laid the foundation for the molecular work of twenty years later.

CHAPTER 13

THE AROMAS OF MIMICRY

. . . plenishing, domestic smells, which compensate for the sharpness of hoar frost with the sweet savour of warm bread, smells lazy and punctual as a village clock, roving and settled, heedless and provident, linen smells, morning smells, pious smells . . .

Marcel Proust, *Swann's Way*

After Philip Sheppard, the most mimicry conscious member of Ford's school was Miriam Rothschild (1908–2005). Officially she was not a member of Oxford University at all, but, as one of the few women Ford ever fully respected, she was very much one of his inner circle. Miriam Rothschild carried the old tradition of the amateur scientist right through into the twenty-first century. 'Amateur' here means amateur in the sense of Robert Boyle, Henry Cavendish, Charles Darwin and other independent scientists with no allegiance to a university or corporate structure of any kind. A member of the illustrious banking family, Rothschild could afford to equip an excellent laboratory in her home at Ashton Wold, near Peterborough. She did have spells in institutional labs, especially the marine Biological Station at Plymouth in the 1930s, but, essentially, she was an independent researcher.

More than anyone else working with butterflies in the twentieth century, Miriam Rothschild combined science and poetry in her approach. Nabokov is the only other contender. The positions are neatly reversed, Nabokov having been an illustrious writer who was also a serious entomologist, Rothschild having been far more of a scientist than Nabokov but deeply imbued with a sense of poetry.

During the Second World War, Rothschild was an Enigma code breaker at Bletchley Park. She had enormous intellectual energy and a boundless love of animals. Her approach to the natural world was questing and multidisciplinary (before this was fashionable), and it is an indication of her originality that she recognised Proust, the arch effete townee, as 'the first and greatest urban naturalist the world has ever known'. Like her, Proust was hypersensitive to odours, moods and the weather.

Rothschild's principal life's work was that humblest of subjects: fleas, in which she become *the* world expert; but she also had a great love of, and a deep insight into, mimicry. Ford was her guide, initially: around Ford, everything to do with butterflies and moths took on a new life. When he came to lunch one day, Rothschild noticed that a frosted orange moth alighting on the carroty hair of her seven-year-old son was almost perfectly camouflaged! The conversation on camouflage and mimicry which followed piqued her interest.

Miriam Rothschild saw mimicry in the round and realised that it was more than a matter of appearances. Her abiding interest was in the way many plants and animals use chemicals to attract and repel, and she recognised the neglected fact that mimicry does not begin with one species copying another. It begins with an insect evolving the ability to harvest and store toxic chemicals from its food plant. Miriam Rothschild was a pioneer of the study of chemical interactions in insect life; for her, chemicals were creaturely characterful substances – almost like plants or butterflies, in fact. She had a special love of pyrazines, some of whose odours you will know even if you've never heard the name before:

Squeeze a ladybird very, very gently, and its characteristic aroma will be on your fingers, for days if you leave it there. That's pyrazines, and there are dozens, perhaps hundreds, of pyrazines, combining to make the aromas of life, from urine, to chocolate, to butterflies, moths and a host of plants. Pyrazines are wonderful, they are universal.

To this pile of pyrazines, I would like to add the aroma of fresh toast. The science of these toxic chemicals and of their role in plants and insects is vast and deeply fascinating. It was the lure of this brand of chemistry that led Miriam to explore one of the most glaring gaps in the understanding of mimicry:

> One of the curious facets of the theories of warning colouration and mimicry is the delay of a century which elapsed between the brilliant and intuitive generalizations put forward by Bates (1862), Müller (1878,1879) and Wallace (1889) and the lab experiments which prove that certain aposematic* insects – the models – sequester and store toxins derived directly from their food plants.

Rothschild set about righting this situation by collaborating with the Polish-born Jewish Nobel-prize winning chemist Tadeusz Reichstein (1897–1996), to investigate the toxic chemicals involved in mimicry. Reichstein, who worked in Switzerland, was a virtuoso natural products chemist. From 1962, Rothschild set out to establish the truth about the toxicity of warningly coloured species – not by tasting them, as Abbott Thayer had done, but by extracting the toxins from the butterflies, identifying them chemically, and feeding them to birds. The model butterfly for this was the American monarch (*Danaus plexippus*), a relative of the African monarch (*Danaus chrysippus*). The toxic monarchs obtain their poison from the milkweed plant (*Asclepias* species). At this point, another researcher, the American Lincoln Brower, comes into the frame. Brower is quite simply the monarch man – he has devoted his life to these magnificent butterflies.

The monarch is a very large, long-lived, strikingly coloured toxic butterfly. It is best known in the popular imagination for its enormous migrations, from eastern North America down to Mexico and back. Like Rothschild, Lincoln Brower wanted to get to the bottom of the business of toxic butterflies, and he supplied her with American monarchs.

---

* Aposematism is the technical term for warning coloration which advertises a creature's toxicity or other dangerous qualities.

Rothschild began to work with a pharmacologist in order to identify the toxins in the monarch. These were suspected of being heart poisons similar to digitalis, the famous foxglove heart drug discovered by William Withering in 1785 and still in use today. Biological tests showed that the monarch extracts did indeed contain substances with digoxin-like properties (cardenolides). Toxicity tests on the extracts were rather startling, causing starlings to vomit.

This was all very promising, but more precise chemistry was needed; and this is where Reichstein came in. He was a chemist who had specialised in the complex molecules of life. Early in his career he had worked on the flavour elements of coffee. These are nitrogenous cyclic chemical compounds, and many of the compounds involved in mimicry are similar. In 1933 he synthesised vitamin C, independently of the team generally credited with the discovery. He worked especially on steroids, of which there are very many with different functions, but all with variations on the same four-ring chemical structure. In 1950 Reichstein and his co-workers won the Nobel Prize in Physiology or Medicine for their work on cortisone, another steroid, elucidating its structure and demonstrating its dramatic therapeutic value in the treatment of rheumatoid arthritis.

Digitalis and the other plant cardenolides have steroid-like structures, and Reichstein was able to show that there were five principal cardenolides in the monarch. This was the first time that a digitalis-like toxin had been found in any animal other than a toad.

Rothschild and Reichstein gave a presentation of their results at a Royal Society Conversazione on 12 May 1966. Reichstein found that both the butterflies and the pupae contain 1.8 times the lethal dose for a cat. But – biology being the science of exceptions – mice are immune to this powerful toxin. Rothschild bred butterflies for the experiments and reported: 'Field mice also raided the greenhouse at Ashton, Peterborough, where the *D. chrysippus* were reared and destroyed 50 percent of our stock of half-grown larvae'. Mice can also eat the milkweed plants, on which the caterpillars feed.

This work made the basis of mimicry much more solid. In fact, mimicry begins even earlier than the harvesting and storing of toxic chemicals. When a butterfly lays its eggs on a leafy plant, the eggs and caterpillars are obviously vulnerable to the herbivores that browse on the plant. Roland Trimen had noticed this way back in the 1860s:

> Apart from the unpalatable nature which renders it distasteful to insect-eaters, there can be no doubt that the wide prevalence of [*Danaus*] *Chrysippus* is largely due to the circumstance that its larva affects chiefly, if not solely, asclepiad plants, which very few, if any, herbivorous mammals will feed upon.

On non-toxic plants, swathes of caterpillars are lost in this way. But, if a butterfly lays eggs on a toxic plant – one that herbivores shun – provided the caterpillars can thrive despite the toxin, the eggs and caterpillars will be sheltered to some degree by the plant's toxic umbrella. So the very first stage of mimicry was a mutation which allowed caterpillars to feed on a plant toxic to most other species. Now the caterpillar has the plant to itself, both for food (nothing else can eat it) and for protection (nothing else comes near it). Once this relationship is established, it is a short step from tolerating a plant's toxins to absorbing them and gaining the same protection for the insect that the plant enjoys. And once protection for the insect is achieved, there is a significant advantage in advertising the fact by way of warning coloration, so that predators don't need to consume too many insects before getting the message. So the bright conventional warning colours follow. As Rothschild put it:

> There seems little doubt that it is the birds' superlative vision which has destined the tobacco-feeding grasshoppers, moths and beetles to be dull in colour and secretive in habit, and the majority of those insects feeding on *Asclepias* and oleander to be gay, eye-catching and self-advertising.

Rothschild was the pioneer in the now burgeoning field of chemical interactions between insects and plants. The prince of contemporary chemical ecologists is Thomas Eisner at Cornell University, who met Rothschild at an insect symposium in 1960 and hailed her as the one 'who introduced the concept of chemical mimicry and probably brought more ideas to the meeting than the rest of us put together'.

Eisner has investigated the chemistry behind countless insect interactions, but the vital spark was created by the bombardier beetle. This
beetle is decked out in black and yellow warning coloration, and what lies
behind the warning is a piece of intricate natural technology.

Eisner discovered the bombardier in Lexington, Massachusetts in
1955, while writing his doctoral thesis. He had been torn between biology
and chemistry. The beetles, when disturbed, ejected a foul smelling spray
accompanied by a popping noise. Eisner had smelt this odour before and
felt he had to know what it was chemically: 'in stumbling on these
little beetles I had struck gold. Bombardier beetles were precisely the
sort of chemical champions I was looking for.' The beetle which Darwin
popped into his mouth, as a student at Cambridge, might have been a
bombardier; it certainly ejected something nauseous.

The spray is a benzoquinone, and the way it is produced was slowly
pieced together in greater and greater detail by Eisner over decades of work.
The spray is hot, almost 100° C, and is delivered in machine-gun bursts at
a frequency of 500–1,000 pulses per second. In action, seen under hi-speed
photography, the beetle looks more like a rocket than a bombardier, and the
propulsion system is uncannily rocket-like too. Two fuels are mixed in a
reaction chamber with enzymes as catalyst. One fuel is hydrogen peroxide,
commonly used in the world of human technology as a rocket fuel; the
other component is hydroquinone. In the reaction, which generates the
heat and explosive power, hydroquinone is oxidised to benzoquinone.
The final touch is that the spray nozzle can be swivelled – gimballed is the
rocketry term – to almost any angle, to hit the attacker directly.

The bombardier beetle's mechanism is such a neat piece of technology that it has been hi-jacked by the intelligent-design lobby. For
them, the mechanism is far too precise and coordinated ever to have
evolved by natural selection. But they gloss over the fact that more primitive bombardiers still exist that, instead of confining the spray to a direct
nozzle, merely ooze benzoquinone froth. The reaction obviously started
in this crude way and has become canalised. Such chanellings of secretions are common in biology. Many male creatures, for instance, do not
direct the sperm directly into the female but release it into the environment, in the vicinity of the eggs. But others do it differently – for instance
the octopus, which has adapted one of its tentacles as a penis.

The lesson of the bombardier is not that such a mechanism casts doubts on evolution, but that in the insect world visual communication is often at the service of chemical interactions. We are such a visual and language-orientated species that it is hard to understand that in the insect world, although sight is very important, chemical communication is often much more so. The warning coloration of the bombardier is secondary to its immensely powerful system of chemical protection.

Poison-spraying beetles are obviously very powerfully protected, and one such species has attracted a mimic of an unusual kind. Batesian mimicry usually operates between quite similar creatures – most often a butterfly copies another butterfly; but flies copy wasps, spiders even copy ants. In theory, Batesian mimicry could operate between totally dissimilar species. To take it to an absurd extreme: if it benefited a small, harmless lizard living in harsh desert conditions to mimic a notoriously dangerous beetle, and if that lizard had the genetic apparatus with which to imitate the superficial form of the beetle convincingly enough to deceive possible predators, this improbable mimicry could occur. It would be a wonderful demonstration of Cott's superficiality principle.

This is obviously a preposterous notion: the sort of idea that an over-enthusiastic Abbott Thayer or Edward Poulton might have proposed. But, however unlikely it seems, this mimicry actually occurs in the Kalahari semi-desert of southern Africa. The lizard is *Eremias lugubris* and as an adult it wears pale red tan desert camouflage. But, when it is young and most vulnerable, it is the same size as the poison-spraying oogpister beetle, and the lizard seems to have taken advantage of the chance similarity in size to evolve a similar black-and-white coloration and a stiff, arched back and tail-depressed gait, far more like those of the beetle than of the adult lizard. To cap it all, the young lizard's tail – which might give the game away (beetles don't have tails) – is not black and white but desert tan. Being camouflaged and pressed to the ground, the tail, inevitably, largely disappears, leaving the impression of a black-and-white beetle. So both the warning coloration and the camouflage are on the same animal.

Any one of these traits shared by lizard and beetle might be thought to be a coincidence, but the combination is convincing. When the young lizards have outgrown the beetles, they adopt their new camouflage strategy. They don't have to pretend to be beetles any more.

If lizards can mimic beetles, what about those strange animal forms that some caterpillars seem to adopt? Poulton was ridiculed for his enthusiasm for these tiny snake heads and alligator heads. For some, these resemblances were pure chance, like a fleeting cloud which has the profile of an animal or a well-known face, but they were fixed by evolutionary genetical chance in the form of the caterpillar. What possible use could it be for a tiny creature to imitate the appearance of a much larger fearsome one? What kind of predator would be fooled?

Rothschild had the answer: birds. She pointed out that birds are nervous predators and will fly off when startled. Anything that causes a bird to do this can give the insect time to escape. As she says, a creature does not have to look exactly like another one to frighten a bird; it needs to evoke for an instant the memory of that creature. Hence the fact that the fear reaction most animals have towards wasps is enough to allow many harmless black-and-yellow striped insects to gain protection from that generalised fright. She called this ability of schematic forms to produce a startle response 'Aide-memoire mimicry'.

Miriam Rothschild had a very clear way of thinking about mimicry, as she expounded to Ford in a letter in 1974:

> The colour of day flying Lepidoptera is imposed all over the world by animals with colour vision. Convergence of colour patterns/schemes is due mainly to two factors: a) selection by bird predators b) selection by male butterflies. . . . I now see mimicry as the result of three sets of circumstances. Heredity . . . worked on by colour vision in birds or insects, with special preadaptations to make the different sorts of mimicry possible.

The point about pre-adaptations recognises that most butterflies are *not* mimetic because they do not possess the necessary pre-adaptations. This is the idea which would be confirmed mathematically by the Charlesworths one year later. And the relationship between the selection of colour patterns for mimicry and the sexual selection of the same patterns is at the heart of the latest butterfly research. Rothschild's paragraph brilliantly encapsulates the whole question. Packed within it is the outline of an enormous programme still going on, designed to understand the three factors and their interactions.

Ford inspired Rothschild, but she was far bolder and more wide-ranging. She was intensely curious and had no entrenched position to defend. Ford, on the other hand, fought for the discipline he had created, ecological genetics, against all pretenders. Ford's narrow view of his discipline reminds me of the quip about the nineteenth-century Oxonian, the Master of Balliol, Benjamin Jowett: 'I am the master of this college / What I don't know isn't knowledge'. For Ford, 'not-knowledge' included molecular genetics, which most scientists recognised as the future of biology. In a letter to his acolyte Philip Sheppard he gently admonished him for (in a lecture) 'handing a bit too much to the students of molecular genetics . . . too much genetic research is being put into it'. This was June 1968, three years after the deciphering of the genetic code had ushered in the golden age of molecular biology.

Ford's detractors had their revenge with the publication of his *Festschrift* volume in 1971. This was a warm occasion, with Miriam Rothschild giving a party and with contributions by most of Ford's mentees (except Sheppard, who excused himself because he didn't have anything suitable – he wrote to Ford saying that he knew he was one of the few people who would understand this strange omission). But the leading British science journal *Nature* gave the book for review to the American molecular biologist Richard Lewontin. The long piece which resulted is an astonishing attack on supposedly effete English mores by a fierce American Marxist. Lewontin suggested that the direction English evolutionary studies had taken after Darwin (admittedly an Englishman but a serious biologist even to Lewontin) was crippled by English class gentility:

In very large part this [Ford's discipline of ecological genetics] has been a British pastime, traceable to the fascination with birds and gardens, butterflies and snails that was characteristic of the prewar upper middle classes from which so many British scientists came. E. B. Ford, the social and scientific quintessence of that tradition, was one of the earliest to devote his attention to the demonstration of natural selection. *Ecological Genetics and Evolution* is his Festschrift and it is replete with primroses, snails, ladybirds and the Pale Brindled Beauty Moth.

Does this mean that 'primroses, snails, ladybirds and the Pale Brindled Beauty Moth' do not obey the laws of biology? Didn't Ford's school do *any* good science? Are butterflies and moths to be damned as subjects for experiment because they, in their innocent way, live in England as well as in other countries? Was Bates wrong to think that evolution might be written more clearly on a butterfly's wings than anywhere else? Since science is judged by criteria that pay no attention to the personal attributes of the researchers, how could Lewontin's distaste for Ford's social niche have anything to do with the worth of Ford's science? Ford's papers contain a penciled riposte to Lewontin, complaining of his 'presumptive personal remarks'.

Ford's ghost continues to haunt entomology. He was an anti-hero in Judith Hooper's book on the peppered moth, *Of Moths and Men*. He presents an easy target. Lewontin returned to the fray in his 1974 book *The Genetic Basis of Evolutionary Change*, and for Hooper Ford's Oxford school were 'silly toffs with butterfly nets'.

But, despite the abuse, both Haldane and Lewontin recognised the excellence of Clarke and Sheppard's work:

> There is no doubt that the work of Clarke and Sheppard has vindicated most of Ford's theories put forward in 1936 . . . a beautiful example of what geneticists can do to explain an apparently rather chaotic situation.
> (Haldane)

> There have been great successes, as for example the case of Müllerian and Batesian mimicry so brilliantly worked out by the Browers and Philip Sheppard. . . . Natural selection not only operates, but works to make tasty butterflies look like nasty ones and nasty ones to look like each other.
> (Lewontin)

Given that Lewontin, Haldane and Ford himself were all agreed on what the Ford school's successes were, it is hard to see what Lewontin's quarrel was, other than *ad hominem* prejudice. Admittedly, Ford was too narrow in resisting the tide of molecular biology, and the sequel to Clarke and Sheppard's work, continuing strongly today, adds molecular biology to all the techniques Ford would have recognised.

CHAPTER 14

THE TINKERER'S PALETTE

God is just another artist. He has no real style. He just keeps trying new things.

> Pablo Picasso, quoted in Françoise
> Gilot, *Life with Picasso*

At some point, after marvelling at the patterns nature is able to copy from one creature to another, it is natural to wonder about the processes which produced these resemblances. Science is the art of the soluble and, until recently, how living patterns are created, how they vary, and how they are inherited were puzzles for which the essential questions couldn't even be framed. For Darwin himself, despite the evidence that he had carefully weighed over the twenty years between his Galapagos trip and publishing *On the Origin of Species*, the speculative nature of his idea and the intangibility of the fleshy processes which created living, evolving things remained a conundrum.

Natural selection is not a thing, an entity, or even a consistent measurable force, as is the force of gravity. The human mind is so constructed that, since that time when names were first applied to things, we have assumed that there is always some *object* that lies behind any name we are offered. As the physicist Richard Feynman put it in saluting his father's role in his education, 'he didn't give me a name; he knew the difference between knowing something and knowing the name of something'. This ought to be borne in mind every time that evolution and natural selection are discussed: the illusion that a name necessarily conjures up an object or some kind of explanation.

From the struggle to live long enough to mate and to produce at least enough viable offspring to replace the parental generation, we infer that the struggle for life has preferred some forms over others. Natural selection is an inference from the different populations of creatures which have existed over enormous spans of time.

The abstract nature of Darwin's conception left out the processes of life – the chemical and physical interactions, about which nothing was known in his time. Living matter is not some amorphous putty. Darwin had no idea what living tissue was made of, but we do know a great deal about DNA, proteins, genes and the other components of the living cell. Darwin's recipe – take some unknown Ur-stuff and subject it to natural selection for 3.5 billion years;* result: the whole of creation – lacks something. Evolution isn't an abstract process. It began in a world which had oceans, mountains, other land forms and clouds. And, once it started, each created life-form interacted both with the environment and with other life-forms. And the life-forms also had their own chemical processes: DNA and proteins create patterns out of the purely chemical logic of their interactions, and natural selection acts on them.

Tangible knowledge is more interesting than the idea of variation and evolution in the abstract. The gallery of mimetic creatures begs an answer to the question: can the new knowledge of molecular genetics explain how all this came about? DNA famously has a forensic value in human affairs, revealing patterns of parentage, crime and human lineage. It can also help us to reconstruct the course of evolution. What are the genetic changes that flip the pattern of a *Papilio* female from one copy of a toxic model to another? Which genes are knocked out to produce the *Kallima* leaf pattern from the Nymphalid groundplan? How are the form, colouring and behaviour of the flower mantis, or the bird-dropping mimicking spider, or a living stone created, and how do they come to reproduce the pattern of an object totally remote from them in material constitution? A butterfly has

---

* Darwin, of course, did not know that life was 3.5 billion years old; this version of his recipe updates his own.

nothing to do with a leaf, a living stone nothing to do with a mineral stone, a spider nothing to do with the relatively formless mash of cells expelled by a bird as waste. Only superficial appearances unite them. But what deep genetic processes create these surface appearances, which are the making or undoing of the creature? And did the patterns we see evolve once only, or have they evolved several times, and been lost and refound? Are there, as some have suspected, ancestral patterns lurking in the genome of creatures, waiting to re-emerge when an ecological need and some serendipitous mutation coincide?

The science of pattern formation in biology is known as evolutionary developmental biology – Evo Devo for short. So much attention has been recently paid, and rightly, to the work of Darwin himself and to his wonderful idea of 150 years ago; but the new biological synthesis, the Evo-Devo revolution taking place right now, is a relatively well-kept secret.

Evo Devo derives from Watson and Crick's epochal discovery of the structure of DNA in 1953 and from the subsequent deciphering of the genetic code by 1965; but these achievements did not initially tell us much about the making of such things as body plans and wing patterns. Watson and Crick showed that the structure of DNA could do two vital things: in its combinations of just four bases it could carry the code to make all the proteins (and, through them, all the other chemicals) which are needed to create and maintain a living creature; and, through the untwining of the double-stranded helix into two complementary strands, it could account for the ability of living things to reproduce themselves accurately. But how these proteins came together to create cells, which in turn grouped themselves to form organs, the whole somehow coming together to make the messy, smelly, panting and heaving creatures that eat, fight, excrete and reproduce themselves – this was completely unknown.

If the main function of DNA is to make individual proteins, what could give the instruction to make an arm, or a leg, or eyes, or a copy of another creature's markings – structures made from billions of cells, each containing thousands of proteins? Where are the big switches that assemble these finely detailed mechanisms into a coordinated whole? And there were some puzzles about the 'beads on a string', as genes were then visualised – stretches of DNA along the chromosome. For one thing, every cell in a creature contains the whole kit of genes; this explains how a new being can be created from a single cloned cell from any part of the body. Why, then, are we and the other creatures not just a formless protein stew, with every gene constantly spewing out the proteins it cannot help but make? The first cell, the fertilised egg cell, obviously needs to contain all the genes a creature would require; but why should an eye cell contain the same genes as a blood cell? How does a cell know whether to be an eye cell or a blood cell, or a butterfly wing cell, and how to cooperate with all the other cells to maintain the organs of the whole body?

For a long time, no answer was forthcoming. Certainly the picture of life after Watson and Crick was far more concerted and detailed than anything previously available, yet of the process of the differentiation of cells there was no sign. But we could guess what had to be happening. The fact that eyes are eyes, nails are nails and nerves are nerves means that in any such part of the body most of the genes *must be switched off.* If they weren't, the organs would dissolve into that undifferentiated stew of cells. This does sometimes happen: it is called cancer.

So what tells the genes when and where to switch on or off? Surely, if genes are the ultimate code to make a creature, there can't be another master code lurking in the background, telling *them* what to do?

There isn't. The DNA inside a living cell has no option but to continue the dance of life, spurred by the living cellular environment in which it finds itself: every stage of its switching on and off is triggered by the stage before. It is now possible to 'stain' genes with fluorescent substances and watch the resulting coloured patterns of gene expression as the genes are switched on and off. Over time, banded patterns appear and disappear – cascades of purposeful body-building gene action.

Think of those elaborate toppling domino patterns so beloved of TV advertisers, in which not just dominos, but books and objects of all kinds, once nudged by the preceding item, cannot help falling and passing the lapsing motion on, to the bitter end. Genes just can't help turning each other on and off until the adult form of a creature is achieved. Except that, sometimes, the machinery gets blocked. I hatched some ermine moths and half the pupae failed to divulge their adult moths. I kept them for a whole year, during which time they neither developed nor decayed, in the hope that, at the due time one year later, on an environmental cue, the machinery would start up again. It didn't. The dominos had become locked into a tangle somewhere. In them the 3.5 billion year clock had stopped, and whatever particular genetic inheritance they had is lost forever.

What is it that kicks the dominos off? When foreign DNA is to be inserted into an egg of, say, a fruit fly, to transform it, it has to be inserted into the correct, posterior part of the egg, because its first actions depend on the existing machinery in that particular part of the cell. And the mother's genes are also expressed in that first fertilised cell before the new genetic makeup of the offspring kicks in.

From this asymmetric beginning, once cells start dividing, they express different genes, depending on their position. This process, creating a complex body plan by first dividing into front, back, top and bottom and then refining each part, stage by stage, is reminiscent of the drawing strategy which creates an animal by first pencilling in three or four ovals which define the outer envelope and then gradually refines the lines of the body, limbs and head within this rough envelope. This is very unlike the way we make, say, a Second World War Sunshield. The Sunshield and a *Kallima* butterfly are both mimetic creatures, but the Sunshield is made by a rigid assembly process. This gradual hatching in of detail from an initial splodge explains so much about life-forms. As the contemporary poet Anne Stevenson puts it in 'The spirit is too blunt an instrument', life begins from the vaguest of blundering gestures:

> The spirit is too blunt an instrument
> to have made this baby.
> Nothing so unskilful as human passions
> could have managed the intricate
> Exacting particulars . . .
>
> Observe the distinct eyelashes and sharp crescent
> Fingernails, the shell-like complexity
> of the ear . . .

Genes are the most paradoxical entities. On the one hand, they display remarkable constancy by comparison with language and culture. For example, think of the game of Chinese whispers. You tell someone 'I'm getting pissed off. I can't stand it anymore!' and ask them to pass it on. After very few relays the message has become something like: 'Armageddon pierstaff! Arcane standards. Hannah More!' (apologies to Kingsley Amis). In 1,000 years the English language has gone from '*Syle us todæg urne daeghwamlican half* to 'Hey man, what's the shit that guy's laying down?' But the genes that keep us ticking over – the 500 or so house-keeping genes in every living cell – have stayed the same since the evolution of single-celled life-forms over 3 billion years ago.

So the genes show a constancy which is hard to believe, over unimaginable stretches of time. Yet they are also, in other respects, slippery will 'o the wisps.* Every human being has small variations in all of his or her genes compared to any other person. There is no such thing as the standard, Platonic form of the human genome. In 2007 Craig Venter, the maverick biologist who raced the official world governmental teams to sequence the human genome (the result was a draw), was the first person on the planet to have his own genome sequenced. DNA pioneer and sometime adversary of Venter, Jim Watson, was the second. Others are following and we are now learning the true extent of human genetic variability, with every gene displaying minor variations between any two individuals. Any two humans share only about 95 per cent of their gene

---

* Just how slippery is becoming more and more apparent. Mobile genetic elements, first discovered by Barbara McLintock in the 1940s and now called transposons or jumping genes, have remarkable properties. Many of these jumping genes have come from viruses known as retroviruses, which insert themselves into the genome of the host. This is not Lamarckian inheritance of acquired characteristics but it does involve extraneous genetic elements entering the germ line. About 8 per cent of the human genome consist of such transposons. A good example is the P-element, a transposon that appeared in the fruit fly *Drosophila melanogaster* in mid-twentieth century and has now spread through all the wild populations of *D. melanogaster*. Originally the transposon had a lethal effect, but it has become neutered. As the evolutionary geneticist Gabriel Dover says: 'Essentially, natural selection has promoted the coevolution of systems of repression of jumping genes ... different repression systems become established in different parts of the world almost, not quite, on a first come first served basis.' A key aspect of the behaviour of transposons is that they clone themselves; when they move they leave a copy of themselves behind. This is the genius of DNA – it abhors gaps, and the complementary coding of its twin helixes makes copying easy. Imagine a page of text on a computer screen. A word suddenly flies out of one line and inserts itself randomly into another. Where it was, the word reforms itself, letter by letter. This happens repeatedly. Transposons are a newly discovered genetic device that may be involved in pattern formation. In jumping around the chromosome, a transposon might land in a place where it can be used as a new control gene. These self-cloning transposable elements have leverage, multiplying their effects many times over, offering new patterns for selection.

sequences. This percentage is five times higher than what was expected and makes, in a sense, two humans less like each other than an average human and an average chimpanzee.

It would be wrong to misinterpret this fact. It happens because most genes can fulfil exactly the same function, even if they have minor substitutions in their genetic code. On the other hand, a single substitution in the wrong place can cause serious diseases. Sickle-cell anaemia is one such. In this condition, the gene coding for haemoglobin, the oxygen-carrying pigment in the blood, carries a mutation. The mutation has a high incidence in African populations because it confers resistance to malaria.

Darwin would have been delighted by the extent of variation now revealed, because variation is the raw material of evolution. This two-faced nature of genes – constant in some respects, extraordinarily variable in others – is one of the keys to pattern forming. The same genes have been discovered in creatures as different as the fruit fly and the human being, but they have often been co-opted to do totally different jobs. Nature is a *bricoleur* – or a 'tinkerer', as the Evo Devo researcher Sean Carroll calls it: not creating genes for new functions from scratch but always using existing genes and adapting them, though mutations, to new tasks.

Thanks to our ability to sequence entire genomes (the total DNA for any organism), first achieved for the fruit fly in 2000, it is now possible to match changes in genes with changes in form, even if we don't yet understand all the stages in between. Some of the ways in which tinkering works are coming to light.

The question of the evolution of mimicry is no different from that of evolution in general, but it is a very particular case, which, if solved, would clear up some of the biggest mysteries of all. As one of the pioneers of mimicry, Poulton, put it, 'The hypothesis that explains mimetic resemblance explains evolution'.

So how does a single supergene in *Papilio dardanus* produce, from one brood, its several different forms of the female butterflies? Conventional genetic work had reached an impasse. The resolution of the problem came with the Evo Devo revolution. What is different now is that we are no longer talking about small or large leaps in something no one has ever seen. We have the base-by-DNA-base total genomic structure of some creatures, and the genes and their functions are being mapped. In the original Watson–Crick model it was thought that the proteins coded for by DNA would account for the differences between a fruit fly, a human and a blue whale – to travel along the huge size range that animals exhibit. But the genes of all mammals are very similar (99 per cent of human genes are found in the mouse, for example). Even insects and humans have many genes in common.

So what accounts for the differences? We now know that it is where and when a gene acts, rather than what it is, that determines bodily forms. For instance, what makes your hands different from your feet is not different genes, but the same genes acting in different locations, in different combinations and at different times. Goldschmidt's insight of seventy years ago has been amply confirmed. A gene is not a fixed instruction to 'make this part of the creature'.

What tells the gene when and where to act is sections of DNA which sit next to the actual protein-coding regions of the gene and either enhance or repress the expression of the gene concerned. These enhancers and repressors are themselves switched on or off by proteins called transcription factors, made by other genes. Consider the skeletons of a human being and a chimpanzee. They have the same number of bones in the same positions, but their relative lengths and thicknesses make for one or the other characteristic shape: the shambling chimp with its arms dangling to the floor; the enormously extended leg bones of the human and correspondingly much shorter arms. It is different timing of gene action that accounts for the morphing of one into the other.

Sometimes, whether a gene is active depends on the summation of many transcription factor inputs. In this, genes begin to look like the networks of our nervous system, in which whether a neuron fires to connect with another depends on the sum of its input signals. The result is cascades of gene action. Another analogy is electronic networks. The genes are a vast network of control elements, constantly computing whether to increase or decrease their output, to make, for example, the shorter or longer bones of chimps and man. In this sense, evolution is a process a little like the way a child's brain becomes connected up by experience. For experience, read natural selection.

Many of the principles of pattern formation can be seen in biology's old warhorse, the fruit fly – the insect which changed the course of biology back in 1910. The fruit fly's entire genome was sequenced in

2000, just before the human genome; but long before that, from the 1980s on, most of the instructions for building a fruit fly had been worked out. In the early 1990s Sean Carroll at the University of Wisconsin-Madison set out to investigate butterfly wing patterns by using the vast knowledge gleaned from the fruit fly.

Carroll is one of the leading lights of Evo Devo. He is passionate about the new biology, defending evolution against the tide of creationism. He is one of those scientists who believe in taking their findings out into the world and publishing popular science books such as *Endless Forms Most Beautiful* (2005), the best account of Evo Devo so far, and *The Making of the Fittest* (2007), an eloquent celebration of Darwinian evolution.

A major finding of Evo Devo is that genes can be co-opted for completely different functions, so Carroll reasoned that some fruit fly genes might show up in novel places, such as butterfly wing patterns. Carroll's team was looking at the expression of a gene called *distal-less* (*dll*), which, in the fruit fly, controls the wing margin. It is called *distal-less* because a mutation in the gene causes the outer (distal) portion of the wing to be lost. *Dll* was found to be expressed in the wing margin of butterflies as well, doing the same job as it does in the fruit fly. But Carroll's team discovered that *dll* was *also* expressed right at the centre of the eyespots of the *Precis coenia* butterfly. They had scored a bull's-eye: the first butterfly pattern gene! An old gene had found a new use again. Nature the tinkerer.

Carroll's team also showed that the basic unit of butterfly wing patterns, the wing scale, is genetically very similar to the tiny hairs on a fruit fly's wing. The wing scale, of course, is the building block in the house of mimicry. Single wing cells have only one colour, so all of these patterns are mosaics of differently coloured scales. This raised hopes that other links between fly genes and butterfly patterns would soon be found. *Dll* is only responsible for the centre of the eyespot. Other fly genes were soon found in other parts of the eyespot.

While the eyespot patterns were being investigated, Sean Carroll and other researchers tried to find relatives of the fruit fly pattern genes in other butterflies, especially in *Heliconius*. This attempt was at the time unsuccessful, suggesting, as had long been suspected, that *Heliconius* is not a Nymphalid groundplan butterfly like *Precis* – its pattern mechanism seems to be distinctive.

Carroll realised that deciphering the mechanism of more intricate butterfly wing patterns was going to have to await advances in Evo Devo techniques, so he resolved to find out as much as possible about the wing patterns of fruit flies first.

In the colourful display stakes, fruit flies are poor things compared to butterflies; but they, too, have their patterns. Carroll addressed the coloration of a species of fruit fly which uses a black spot on its outer wing margin in courtship (it is intriguing that the apparently workaday, quick-breeding fruit fly goes in for courtship rituals at all). These black wingspots developed by gaining new binding sites on the wing margin for the expression of the black pigment. How did this happen? In the present case, new binding sites evolved for regulatory proteins in a pre-existing regulatory element. But the plasticity of regulatory systems can go way beyond this. Carroll has discovered that, if you disable a regulatory gene (molecular biologists have sophisticated techniques for knocking out specific genes), you often find that a piece of inactive, disregarded, junk DNA steps into the breach and begins to function successfully as a control gene. This is as if, in a soccer match, when a player is injured he is replaced not from the bench but by a paunchy unfit guy plucked at random from the terraces and – *mirabile dictu* – he does a perfectly reasonable job. This idea of cryptic, that is, hidden, genes is a powerful one: it seems that pattern-forming potential lurks in such hidden genes, presently doing nothing useful.

Nature scavenges genes and adapts them to new purposes rather in the way the Second World War *camoufleurs* scavenged waste dumps in the desert for materials The reason why this is possible is that the action of genes is entirely chemical. Chemicals usually have a range of functions and so it is possible with very small changes to a chemical molecule to adapt it to a new purpose. A brilliant example of gene scavenging is found in Antarctic fishes. The current Antarctic ice did not appear until around 10–14 million years ago (it had been frozen in previous eras). To cope with living in waters permanently below freezing, fish either had to develop a form of anti-freeze or perish. The gene that produces the anti-freeze has been traced to a variation on a gene which produces the digestive enzyme trypsinogen. The gene has been copied and mutated to create the new 'anti-freeze' gene. This is one method evolution has used to develop new functions: an existing gene is duplicated and, because the copy is not needed to fulfil its old function, the copy can mutate to do a different job.

Another fruit fly research team, in 2007, both solved the old Darwinian small cumulative variations versus large discontinuous leaps dispute and answered the puzzle posed by the Charlesworths in 1975. David Stern's team at Princeton has worked on a different aspect of pattern formation in fruit flies. The bristles on the wings come in various patterns – crudely put, hairy and bald. (Remember that fruit fly bristles are the evolutionary forerunners of butterfly wing scales.) Stern's team has tracked the genes responsible for the loss of bristles. The regulatory gene concerned is known as *shaven baby* (*Svb*). Their paper in *Nature* in August 2007 reported:

> *Svb* [the shaven baby gene] seems peculiar in the network of genetic interactions that establish the bristle pattern because it sits at the nexus of the upstream patterning genes and the downstream effector genes.

This means that, being poised at this junction, the *Svb* genes can make big changes in the desired area without changing other parts of the body. Here is the answer both to the hopeful monster problem and to that of pre-adaptation (namely that the genes involved in complex mimicry need to be already linked *before* they develop mimetic patterns). Conventional Neo-Darwinists claimed that monsters were never hopeful, that large mutations would always be damaging. Many genes have multiple effects – so with a hopeful monster you might obtain your mimetic wing pattern, but the wings might not flap, or the butterfly might be sterile or blind. But a gene located just at this critical point, with its battery of modifier genes deciding where and when it is switched on, can make this big 'hopeful' change without damaging the rest of the system.

Stern's team says that the *shaven baby* gene has to be in the position it is in because big changes before this point would produce those multiple damaging effects. They go on to reach a conclusion of enormous consequence for the theory of evolution: 'Our results provide experimental evidence that the conflicting views of micromutationism and macromutationism can actually reflect observations of the same molecular mechanisms at different levels of resolution.' This resolves the argument which has raged through this book, going back to the time of Bateson and de Vries and Punnett in the early 1900s, through Goldschmidt and Fisher in the 1930s and '40s to Ford in the '50s and Dawkins and Jay Gould in the '80s: it answers the Big Endian and Little Endian theories of evolution. *Both are true.* What seems a big leap (a macromutation) in the organism itself is the result of a small change (micromutation) in the patterns of gene *regulation* – precisely as Richard Goldschmidt had predicted.

Sean Carroll and David Stern's experiments are models of what we should like to see carried out in the pattern-forming genes of the mimicking butterflies, but this is going to be harder. There are several reasons for the awkwardness of working with butterflies. A fruit fly has only four chromosomes; a butterfly has twenty-one. Butterfly breeding is much trickier and lengthier than fruit fly breeding, and no butterfly's full genome has yet been sequenced. A bid to do this, made in 2004, foundered because the butterfly research community was divided into advocates for one out of two species of butterfly to be sequenced. Most of the recent work has been conducted on *Bicyclus anynana* and *Heliconius*. This was a difficult choice: *Bicyclus* has produced results with its eyespot genes, but it is not a mimic: it is *Heliconius* that holds out the promise of solving the mystery of mimicry, so the butterfly research community applied for funding for both research projects and was rejected. Since then, the cost of genome sequencing has been falling drastically. The human genome project was virtually a moonshot, involving international governmental collaboration. But, as is the way with these things, progress so often makes a mockery of the early heroic efforts.

Genome sequencing is following the trail blazed by computing power. It doesn't yet have its Moore's Law (computing power has doubled roughly every two years since 1968), but it probably soon will, thanks to a new sequencing technology called 454 sequencing. The goal is the $1,000 genome sequence. In 2008 the butterfly community suddenly realised the potential of this project, and there are three butterfly sequences currently under way: *Heliconius, Bicyclus* and the monarch, *Danaus chrysippus.*

What would we really like to do with the butterflies once we have all the genetic tools required for the task? The clue comes from a dramatic piece of work with sticklebacks. The stickleback uses armour rather than camouflage as its defence: it is an 8th Army tank rather than a dummy railhead. The ocean-going form has between 32 and 36 armoured plates running along the sides from front to back. But the sticklebacks have also repeatedly evolved freshwater forms, in lakes and streams created through the melting of the ice during the last glaciation, between 10,000 and 20,000 years ago. Here, it seems, the heavy armour is a liability, and in every freshwater form most of the armour has been lost. This must have happened independently dozens of times. Several explanations for the loss of the plates have been advanced; whatever the real reason, since all freshwater forms have adopted armour-lite, its adaptive advantage must be clear.

In 2005 David Kingsley's team at Stanford University School of Medicine located the genes involved by means of a classical technique, which will probably be used many times again to find such genes. The marine and freshwater forms can still interbreed, and from such crosses a linkage map was created in the time-honoured Thomas Hunt Morgan fashion (these 'old genetics' techniques are still essential in the DNA era). This map showed that the change of armour could be traced to a single gene, which exists in two forms. In other words the cause of the change is a polymorphism such as we have seen in *Papilio* butterflies.

Molecular biology techniques were then able to locate the DNA sequence concerned. It was found that the armour-lite form of the gene exists all the time at a low level in the marine population (3.6 per cent). In that environment it is recessive, so fish with one copy of the gene appear with normal armour; fish with two copies of armour-lite will not survive long in the ocean and will not pass their genes on. But in fresh water the armour-lite version quickly spreads throughout the population, until the whole population consists of the lite form.

How can we be sure of this? The important test is to restore the full armour by inserting the gene for complete armour, through genetic engineering, into the armour-lite embryo. The way Kingsley's team accomplished this task highlights one of the most remarkable aspects of Evo Devo. The gene they located as being responsible for the armour plating (*Ectodysplasin – Eda*) is found not only in sticklebacks: it is a well-known signalling gene in all mammals, including humans, with a range of effects on skin, scales, teeth, sweat glands (once again, the principle is: the same gene / different timing and place of action). Kingsley's team inserted a mouse *Eda* gene into stickleback embryos. Additional plates grew in some of the transformed sticklebacks. This proved the point that *Eda* is certainly a major factor in the presence or absence of plates; but the transformed fishes were not perfect replicas of the marine fishes. Other refining factors may be present, or it may simply be that the mouse gene, although effective, could not replicate precisely the effect of the natural fish gene. It is remarkable that it worked at all – a crude spanner thrown into the works, nevertheless subtly tweaking the mechanism to produce armour plating.

As Kingsley's paper points out, 'Our results show that large changes in vertebrate skeletal morphology in the wild can be created by relatively simple genetic mechanisms'. Yet again, here is a link between a small genetic change and a large physical change in the organism. The stickleback approach should work for very many other adaptations, including mimicry and camouflage.

So the goal is to find the genes which will switch a butterfly wing pattern from mimicking one butterfly model to mimicking another. In this way we can trace the likely evolutionary changes which produced mimicry. First, we need to find the butterfly mimicry genes. The hunt is on, and the subject is the *Heliconius* butterflies of South America.

CHAPTER 15

THE *HELICONIUS* VARIATIONS

Such colours can't be drawn
        from non-existence.
Tell me, at whose insistence
        were yours laid on?
Since I'm a mumbling heap
        of words, not pigments,
how could your hues be figments
        of my conceit?

Joseph Brodsky, 'The Butterfly'

Heliconius butterflies are Müllerian mimics: that is, they are different species wearing the same patterning – often several species grouped together in a mimicry ring – to give a standardised warning to predators that they are unpalatable. Their wing patterns are complicated: they are not geometrically regular, but have splashes of colour thrown on with artistic abandon. Think Matisse's *Jazz* cut-outs, with their bold yellow and black, white and blue – and red. The biologist Sir Alister Hardy (1896–1985), who was a *camoufleur* in the First World War, made a strong claim for butterfly wing patterns being a form of art:

I think it likely that there are no finer galleries of abstract art than the cabinet drawers of the tropical butterfly collector. Each 'work' is a symbol, if I must not say of emotion, then of vivid life; they are arresting signs to warn hungry hunting predators, vertebrates like ourselves, of danger, or they are glowing courtship colours flaunted by male insects to attract and coax coy mates to submission. It is often, I believe, the fascination of this abstract colour and design, as much as an interest in biology or a love of nature, that allures the ardent lepidopterist, although all these may be combined; he has his favourite genera and dotes upon his different species of *Vanessa* and *Parnassius*, as the modernist does upon his examples of Matisse or Ben Nicholson.

These *Heliconius* patterns are hard to describe, but happily the butterfly community has come up with some cute slangy names for some of them. The first and easiest is the tiger-stripe pattern. *Heliconius numata*, a leading player in our search, belongs to this vast group of mimicking species. The tiger stripes are, as you might expect, yellowish-orange and brown. *Numata* is highly polymorphic, with up to seven different forms participating in different mimicry rings in the same location.

Then there are the postmen. The postman *Heliconius* has red forewing bars and sometimes yellow hindwing bands on a black background. The postman appellation comes from the colours of the Trinidad postal service. The classic postmen are *Heliconius melpomene* and *H. erato*, with their thirty different but always paired patterns: black with red splashes, sometimes rayed and sometimes with yellow bands. From place to place, *erato* shows differing wing patterns; but, whatever the pattern, *melpomene* always faithfully copies it. It is thought that *melpomene* is the one mimicking *erato* rather than the other way round, *erato* being an older and more numerous species than *melpomene*.

Some of the postmen patterns are called Dennis, after Dennis the Menace. Dennis the Menace was originally a single butterfly in a brood hatched in breeding experiments at William Beebe's (1877–1962) research station at Simla, Trinidad, the hotspot for *Heliconius* research in the 1950s. The story goes that it was named after the US cartoon character and not after the British kid in the *Beano*. This seems odd in that the US Dennis is rather tame compared to Dennis the Menace from the *Beano*, whose appearance screams 'I am toxic'. Nevertheless, the red patch on the black forewing often seen in *Heliconius melpomene* is 'Dennis', and the beautiful striped rays on some hindwings are 'Dennis-rayed'.

Heliconius cydno is more restrained, in black and white – an Op-Art butterfly. It occasionally has traces of red, but its usual variation on the black and white is a splodge of yellow or blue – very Matisse, very *Jazz*.

If the *Heliconius* mimetic wing patterns are cameos of modern art, how are they painted? They have the blend of abandon and control which we always find satisfying in visual art. Patterns more regular than this are too geometrical to be aesthetically pleasing; make them more haphazard, and there's nothing for the eye to get hold of. The wing patterns were certainly not produced on the Jackson Pollock principle of dripping paint excitably over the canvas – something, one assumes, nature would be good at if it had any reason to do it.

The biological painter obviously uses different rules from those that guide the human hand – but where should one look? To the fruit fly, of course, whose patterns are primitive but whose kit of genes – for processes, body forms and patterns – is better known than for any other creature. Remembering that genes are scavenged and put to new uses, it seemed likely that the butterfly pattern genes would be just such fruit fly genes, recruited to new functions in the service of mimicry on the finely painted canvas of a *Heliconius* wing.

Painting is a useful analogy: on butterfly wings, nature is painting with the invisible hand of natural selection. What we know of butterfly wing patterns suggests that two aspects of painting are involved in nature's patterning process.

If you want to make a painting, you can either just paint and let the edge of each brush stroke dictate the shape, or you can make a sketch of the outlines first. If you make a sketch first, then painting, at its crudest, can be a colouring book exercise – just fill in the outlines. That's simple – it's the way kids do it.

If you don't want to make a sketch first, you run into the problem that it's quite hard to know where the paint should stop to create the patterns you want. From the earliest days of the twentieth century, when biologists thought about pattern formation they imagined that washes of chemicals might spread across the wings during development and just stop where they ran out of steam. Or two gradients might meet and form patterns from the way they interacted. But perhaps there was a pre-patterned sketch that the colour merely filled in? A sort of black-and-white in which the areas to be coloured in later were marked out? Until recently, there was no way of finding out.

The attempt to discover how *Heliconius* butterflies have evolved their mimetic patterns has looked at patterns and colours separately. This approach wasn't based on any assumption that the sketch-plus-colouring hypothesis was correct, merely that the genetics of patterns and the chemistry of colour could only be studied separately.

Work on patterning goes back a long way. In the 1960s and '70s Philip Sheppard and John Turner used the classic techniques of Mendel and Morgan to map the pattern shifts between the species. This involved thousands of laborious crossing experiments and the result was to show that just a few genes 'of major effect' switched the patterns of *Heliconius* from one to another. 'Major effect' means what it says: they were not hopeful monsters, but neither were they micromutations. Unlike *Papilio*, there did not seem to be supergenes – one gene controlling the entire wing pattern – in *Heliconius*.

These key genes switched on and off the typical *Heliconius* markings – for instance the forewing red bar, the hindwing yellow band, the red rays. Some of these genes were linked, so that a grouped pattern would recur – for instance the Dennis appearance: red forewing band, hindwing bar and rays.

This was old-fashioned genetics. Nothing was known of the actual genes involved; nothing was known about the DNA. All that could be determined was the chromosome on which each pattern element was located, and what element was linked to what other element.

Enter the modern era, and such crossing experiments are still being conducted. But now it is possible to locate the actual DNA involved in these pattern switches. A large, closely coordinated, worldwide team of *Heliconius* workers is researching every aspect of *Heliconius* – from field experiments setting mimicry in action to the Evo Devo of the pattern genes. At the centre of this network is Chris Jiggins. Jiggins's butterfly odyssey began in Cambridge, where, as a student, he published his first paper with Michael Majerus. He returned to Cambridge as Head of the Butterfly Genetics Group in 2007, with an ambitious research programme. His wife, the Colombian butterfly researcher Margarita Beltran, is also a member of his team. Much of the work of this team has focused on the two postman co-mimics *Heliconius erato* and *H. melpomene*; it builds on Sheppard and Turner's work, but uses molecular markers in crossing experiments to home in on the pattern genes, instead of simply observing the wing patterns. In 2006 the researchers found that the colour-pattern genes in *erato* and *melpomene* mapped to the same sites on the genomes in each species. This was a confirmation of one of Goldschmidt's most hotly disputed ideas: that mimicry might involve the same genetic processes in both model and mimic. E. B. Ford was quite sure that this was wrong, because he knew of cases where the chemical pigments to produce, say, red in model and mimic were different. But clearly there is more than one road to Rome: parallel genetic mechanisms can sometimes be the source of mimetic patterns.

There may be no complete butterfly genome yet, but there are huge databases of gene sequences for butterflies and moths. The silkworm moth has been fully sequenced in both China and Japan and, because so many genes are common to all creatures, the silkworm and butterfly genomes are very close. As raw data, a gene sequence is just a set of coded letters representing the sequence of the bases in the DNA, such as: ATCGGTACCGATATCGCGTAACTGACAGCTACACTG. This is meaningless by itself, but computer programs can match a new sequence against known gene sequences on the same principle by which you can search for a single word in a word processor file, or Google a key word.

At first, everyone was optimistic that those fruit fly genes would show up as butterfly-pattern switches; but, no, the sequences remained blank. The general whereabouts on the chromosomes of the pattern-switching gene are known thanks to Sheppard and Turner's work in the 1960s and '70s, further developed by the current Jiggins team. But only now are they beginning to find real genes in the DNA of the switching region of the genome. The favourite is a kinesin gene. Kinesins, as their name implies, are force-producing protein molecules; they are widely involved in moving things round in the cell. It is not obvious why such a motor protein would be involved in butterfly patterns but, here again, perhaps the tinkerer is at work: a gene has been co-opted for a very different function from its original one.

Totally separate from the search for pattern is the search for the colour genes. Butterfly colour is produced by a mosaic of single cells, each having a single colour. In *Heliconius*, the colours that scales can take up are limited to red, orange, yellow, white or black. The chemistry of these pigments and some of the chemical pathways in living things by which they are formed have been known for a long time, but not the genes which give rise to them. The most cogent account so far has come from Professor Bob Reed's lab at the University of California, Irvine, south of Los Angeles. Reed, a disciple of Sean Carroll of Evo Devo fame, is working with *Heliconius erato*.

As with the pattern genes, Reed looked for genes already recognised in the fruit fly, to see if they have been co-opted for new purposes in the butterfly. In 2005, Reed's team found that the red-eye pigment in fruit flies is also the red pigment splashed so gorgeously across the wings of *Heliconius* butterflies: nature the tinkerer, the recycler – again. This takes us back to Thomas Hunt Morgan's white-eye fly (see p. 64), the mutant which, in 1910, set genetics on a new course. That fly lacked the red pigment gene. In *Heliconius*, nature has gone one step further than the simple mutant which produced the white-eyed fly and found a whole new canvas on which to paint. This particular red pigment hasn't been exploited by all butterflies – as Ford noted, there are others reds as well.

Reed found that the formation of the red scales seemed to require an *overlap* of two genes: *cinnabar* and *vermilion*. This fact led him to surmise that the yellow/red switch in *Heliconius* might be caused by a regulatory gene which controls expression of the *vermilion* gene. If it doesn't overlap with *cinnabar* the result is yellow; where it does overlap, the wing scales are red.

Reed's discovery that overlapping patterns of gene expression could be necessary to create the forewing bar has an important implication for evolution. It means that, sometimes and in some places, genes will be expressed where there is no overlapping, and hence there will be no visible effect. Natural selection will not act on such areas of gene expression: they are hidden (or cryptic). But new patterns could be created by a mutation which suddenly increases or decreases the areas of overlap: then the 'hidden gene' would suddenly come into play.

The cryptic patterns idea is an exciting one. Think of the wing patterns as containing, beneath the visible patterns, other potential patterns, written in the invisible ink of overlapping regions of hidden gene expression, which a mutation or sometimes an environmental cue can switch on.

Following Reed's work, the Jiggins team then set to work on *Heliconius erato*'s co-mimic *H. melpomene*, to see if the same genes produce the same colours in both species. *Heliconius melpomene* mimics *erato*. The scientists are pretty sure it's that way round because the DNA timeline suggests that *melpomene* is a relatively new species – perhaps 1.5 million years old whereas *erato* is about 10 million years old.

The mimicry between them isn't perfect. In the Amazon, races of *erato* with Dennis rays have the rays confined between the wing veins; in *melpomene*, they cross the vein boundaries. The red forewing band, which is cleanly demarcated in *erato*, is sometimes blurry in *melpomene*. Jiggins suggests that the overlap of the *cinnabar* and *vermilion* genes found by Reed does not occur in *H. melpomene* and that this may account for the latter's fuzzier band boundary. The lesson from this, if confirmed, is that, to make precise patterns, two genes are better than one in controlling the forms.

But, beyond the minor differences between *erato* and *melpomene*, the *Heliconius* team wants to put the picture back together, to join up the dots and to show how the pattern-making genes link up with the colour genes so as to tell the colour where to go. At the moment, the two strands of work are separate. Jiggins and Reed believe that the colour genes are what is known as downstream of the pattern genes, that is, they come into play later in development. But wrenching the two aspects of the process together is going to be a task. The spur, though, as in all science, is that we know the answer is there. The pattern genes must interact with the colour genes. But the colour genes do not map to the same genetic sites as the switch genes for pattern. It would have been neat if they were one and the same, but they are not. The missing link is there to be found.

Heliconius melpomene and *Heliconius erato* have converged as co-mimics in recent evolutionary time. Mimicry can also cause *divergence. H. melpomene* is believed to have split off from *H. cydno* about 1.5 million years ago. *Melpomene* is a postman, with plenty of red, *cydno* is a black and white Op-Art butterfly. Despite their very different appearance, they can still interbreed. And the hybrids they can produce are giving us a new understanding of the process of mimicry.

The basic biology of *cydno* and *melpomene* is very similar, and this is revealed in the DNA, especially the mitochondrial DNA.* *Heliconius* butterflies all evolved from a common ancestor. *Cydno* and *melpomene* diverged perhaps 1.5 million years ago; *erato* split off further back, perhaps 10 million years ago. How did *cydno* and *melpomene* come to diverge so much?

The idea is that species of *Heliconius* are 'captured' by mimicry rings, the dominant protection racket in their area. Whatever is the dominant pattern in a particular habitat, natural selection will force a butterfly to try to match it – to become a Müllerian mimic. Remember: Müllerian mimics are species, unpalatable to predators, which share the same warning coloration, thus making it easier for predators to leave them all well alone. This 'capture' is a pivotal business: this is destiny – which way will they swing? *H. melpomene* joined the postman ring and *cydno* the Op-Art ring.

It is natural selection that is calling the shots, but the effect is as if a butterfly were 'choosing' to look like another. The crucial 'decisions' for a *Heliconius* butterfly are whom to mimic and with whom to mate. As Miriam Rothschild put it: 'Convergence of colour patterns/schemes is due mainly to two factors: a) selection by bird predators b) selection by

---

* The fossil record was once the only evidence we had of evolution. But all DNA in the world derives from the first DNA, and although, during the course of evolution, it has been overwritten countless times, like a computer disk with text constantly rewritten over deleted portions, it still carries crucial evidence of its past, particularly since there are two forms of DNA in every organism, which are inherited differently. The evolutionary history of, and relationships between, butterflies can be gleaned from analysis of mitochondrial DNA. Mitochondria are tiny energy-providing substructures inside every cell in every living creature. They have a quasi-autonomous existence, with their own DNA, inherited only in the female line. Their independence is thought to derive from a former, free-living existence as bacteria before they fused with the single-celled organisms that gave rise to all the multi-celled creatures today. The mitochondrial DNA does not take part in the genetic shifts which create new body patterns – it continues independently to accumulate non-functional mutations over time, at a more or less constant rate, which is independent of natural selection. In this it behaves rather like the decaying radioisotopes, which allow us to date the rocks and to give the age of fossils. It is a mitochondrial clock. So changes in mitochondrial DNA can indicate when two species diverged, while comparison of the nuclear DNA can show what the changes were which produced new forms.

male butterflies.' How this plays out with *cydno* and *melpomene* is emerging from the rain forests of Colombia.

In the 1990s some strange findings began to emerge from the Eastern Cordillera of the Andes in Colombia. There a couple of *Heliconius* specimens seemed to be *melpomene* in appearance, but their DNA revealed them to be *cydno* underneath. The *melpomene* appearance was entirely superficial; they were mimics. Now this was odd, because *cydno* and *melpomene* were not known to be mimics, one of the other. If this could be confirmed with more specimens, here was a missing link, if ever there was one!

Why wouldn't you expect them to be mimics? The standard explanation is that the patterns of the two mimicry rings to which they belong are too far apart: there is a conceptual mimicry 'valley' between them. It was thought to be impossible to pass from one peak to the other because any *cydno* that looked a little like a *melpomene* or vice versa would be quickly picked off by predators, being so different from the patterns of its established mimicry ring.

The Colombian butterfly researcher Mauricio Linares at the University of the Andes in Bogotá had a strong conviction that the region held the key to *Heliconius* mimicry and speciation. There was a missing link between *melpomene* and *cydno* to be found here. As a student working his way through college in Bogotá, Mauricio was inspired by a lecture to study the evolution of his country's bounteous butterflies. He achieved this, doing his PhD with Larry Gilbert, doyen of *Heliconius* studies, at the University of Texas, Austin.

Mauricio followed the trail of the elusive *melpomene*-like *cydnos* into the hills south of Villavicencio, Colombia. There he ran into the turbulence in the region. In 2005 he was searching for the 'missing link' near Puerto Rico (a town in Colombia, not to be confused with the country), Caqueta province, when he was suddenly surrounded by the 'Columna Teófilo Forero', a guerilla band which insisted that he should go with them to be interviewed by their commandant, Giovanni, in the deep forest. The guerillas had previously captured hundreds of hostages, so the situation looked very dangerous. Mauricio concocted a story that he had injured his knee and needed medical attention. After a three-hour 'talk', they eventually let him go to his hotel in Puerto Rico, but suggested that he should meet up with them the next day at 8 a.m. He returned to Puerto Rico with his driver, 'drank a whole bottle of *aguardiente* to relax after such an adventure, and left in the first taxi at 4:00 am next morning to Bogotá'.

This slowed the chase down somewhat; but, on a visit to Bogotá in 2007, *Heliconius* researcher Jim Mallet from University College, London was looking at some *Heliconius* specimens collected by a Colombian student on a 1998 expedition to Churumbelos, a region of Colombia where the Andean mountain region meets the Amazon (and where oil, narcotics and terrorism also co-exist). The student was Blanca Huertas, now Curator of Butterflies at the Natural History Museum, London,* and some of the '*melpomenes*' in Blanca's collection aroused Mallet's interest: he thought they might be the missing link: the *melpomene*-like *cydnos*. So he had their mitochondrial DNA analysed. It proved to be of the *cydno* type.

Mauricio stepped up the chase for live samples. Following his tangle with the guerillas, he had arranged for his samples to be sent to Bogotá by his driver, who still lives in the danger zone. A beautiful Dennis-rayed butterfly with red and yellow on the forewings was found near Florencia. In appearance it was obviously a *melpomene*, but the mitochondrial DNA indicated otherwise. The results were published in 2008 as 'Two sisters in the same dress'. It was now clear that the mimicry valley between *melpomene* and *cydno* could be crossed after all: *cydno* can deck itself out with the colours of *melpomene*. But how? Where is the genetic switch? Could it be a hopeful monster mutation? Or perhaps hybridisation, since they can interbreed? Perhaps the *melpomene* colour patterns in *cydno* have come directly from *melpomene*? This was the line the *Heliconius* team decided to investigate.

Another *Heliconius*, long suspected of being a *cydno/melpomene* hybrid, *Heliconius heurippa*, has proved something of a poster boy for the genus, achieving newspaper headlines in 2006 ('New species synthesized in lab'). One of the many variable forms in the region, *Heliconius heurippa* is not

---

* The Churumbelos expedition changed Blanca Huertas's life: she is now the public face of butterflies in Britain as the insects move back to the museum in South Kensington. The new Darwin Centre, which opened in September 2009, is a gesture of faith, 200 years after Darwin's birth and 150 years after the publication of the *Origin*, in the importance of biological research.

itself a mimic, but it has revealed a great deal about how the dazzling range of *Heliconius* butterflies has evolved. Its wing pattern seems to be an obvious cross between the two with, on the forewing, a yellow band from *cydno* and a red one from *melpomene*. In 2001 Chris Jiggins had discovered that, despite being closely related and capable of interbreeding, *cydno* and *melpomene* were isolated by the simple fact that they choose to mate with their own kind on the basis of wing pattern recognition, the males choosing the females. It seems likely that isolation as species was achieved simply by a refusal to mate (most of the time) with the 'wrong' wing patterns. Interestingly, 150 years earlier Darwin had ruminated on the phenomenon of similar races co-existing in the same place and preserving their integrity by '[evincing] a strong predilection to pair together'.

Chris Jiggins, Mauricio Linares and their team decided to test the theory of the hybrid origin of *Heliconius heurippa* by trying to breed it in the lab. By a process of back-crossing to the parental generation (to get round the problem that the first-generation hybrids are sterile amongst themselves), the team was able to create a butterfly identical with *H. heurippa* in the lab in only three generations. In case you are wondering: these backcrosses occur at a low level in the wild too – which is how, we think, *heurippa* came to exist.

The lab-bred *heurippa* was announced to the world with a fanfare in 2006. Now comes the really interesting part. In 2008, Chris Jiggins showed that these 'artificial' butterflies preferred to mate with their own kind, even though, as far as they were concerned, they only evolved yesterday. As Jiggins says, 'that's pretty cool!' That wing pattern is all for these butterflies (odour doesn't come into it) was confirmed by experiments with paper models bearing accurate wing patterns: these were just as attractive as the real thing, which meant that preference for the *heurippa* pattern was genetically inherited *along with the pattern itself*.

In human terms, this is as if the genes which created a particular type of looks – say, a brown-skinned, blue-eyed redhead – also created a preference for those looks. If the child is a girl, she would inherit those looks; if a boy, he would *automatically favour a girl with the same looks*. Of course, this is cheating for the sake of a good analogy. Human sexual preference involves free-will, cultural conditioning and the Lord knows what complications. But in butterflies there is no element of 'choice'.

Their actions are genetically programmed. The *heurippas* have to follow their genetic instructions to choose a mate as laid down in the genes. It is a perfect little system. So what kind of genetic mechanism can it be that links a new pattern with a ready-made preference for that pattern? We don't know, but it's going to be an exciting chase to find it.

This has implications for the formation of new species of mimetic butterflies. If a viable new wing pattern is produced by hybridisation – and viable in this species means that it advertises its toxicity by flashy coloration, usually in mimicry of another flashy butterfly – the population will be isolated from its parents very quickly by this automatic mate preference for the wing patterns, even though interbreeding and dilution of the pattern are still technically possible. There is now a consensus that, in *Heliconius* at least, such hybridisation is one route to new species.

The idea is that *Heliconius* has a tool-kit of pattern genes (shades of Goldschmidt again) and that hybridisation allows genes to pass from one species to another. When a new ecological opportunity arises, an ancient gene can resurface to create the appropriate pattern, perhaps to mimic a new species which has come into the area or to switch mimicry rings when an old species has died out in a particular area.

Of present-day *Heliconius* species, 20 per cent are known to be capable of hybridising and, in this way, of gaining new wing-pattern genes. To inherit ready-made genes for patterns is a far more efficient mechanism than to wait for a mutation. Patterns might be passed back and forth, gained and lost several times during the course of evolution. This idea has been supported by Sean Carroll and other Evo Devo workers, who have shown that, in the fruit fly, this is precisely what has happened.

But if two species can mate and produce a new species through hybridisation, where does that leave our definition of a species? Poulton said that a species was 'a formed reproductive community'. Distantly related species cannot interbreed at all. But *Heliconius* butterflies are thought to have evolved rapidly fairly recently. As Darwin said of Bates's work, we are here seeing evolution in action. In these cases, the barriers between the species have not yet hardened, and mate choice is the means the species use to remain distinct.

But, when mimicry is very accurate, isn't this going to interfere with mate choice? How do you stop two species of Müllerian mimics such as

Heliconius erato and *Heliconius melpomene* from mating with each other? Some species use odour as a mating cue to get round this problem but *erato* and *melpomene* have a drastic strategy of their own, as Miriam Rothschild graphically illustrated. In the case of *erato*,

> The males display a sexual ardour rarely met with in the insect world. They are attracted – probably by some subtle odour – to the chrysalis of the unhatched female. And they sit and wait in deadly quiet for her to struggle free. She is then raped by the most active and strongest of the two or three attendant males. Not infrequently desire gets the better of them and they violate her – or attempt to do so – while she is still within her chrysalis.

Melpomenes, as we have seen, don't do this. They go in for more decorous couplings by male choice, so there is no confusion.

Heliconius cydno, melpomene and *erato* are gradually revealing their secrets but the real joker in the pack is *Heliconius numata*. Sheppard and Turner had already noted in their 1985 paper that *melpomene* and *numata* were closely related and could interbreed. They suggested that *numata* must have diverged in evolution before *melpomene* acquired its red forewing band. When the Jiggins team mapped the switch genes in *melpomene* and *erato* in 2006, they also mapped *Heliconius numata*, with its so different tiger-stripe pattern in orange and brown. Here there was a big surprise: *numata*'s close kinship to *melpomene* was revealed at the molecular level when a gene which turned on the whole pattern mapped to a gene in *melpomene* which controlled only the yellow hindwing bar. So here was a great genetic evolutionary puzzle. How had the gene for the yellow bar in *melpomene* come to control the whole pattern in *numata*?

The Jiggins team has suggested that the huge leap to the tiger pattern in *numata* 'implies an extraordinary "jack-of-all-trades" flexibility'. It is as if your light switch had freakishly got wired into the Christmas illuminations in the high street and turned on the whole shooting match. *Heliconius numata* is reminiscent of the *Papilio* butterfly wing patterns, which are also controlled by a single supergene; but its pattern is even further removed from that of its other morph, *H. melpomene*, than the *Papiliones'* pattern is from theirs.

It is only a matter of time before this powerful pattern gene is unmasked. The next step will be to repeat the stickleback experiment: to clone the *numata* pattern gene and insert it into a *cydno* or *melpomene* eggs. The prediction is that the *numata* pattern will be switched on, just as the stickleback armour is switched on and off by the variant – ocean-going, armoured or freshwater lake, unprotected – form of the gene.

Heliconius butterflies are exciting for many reasons. They are evolving faster than most other creatures on the planet; they have the *H. numata* supergene; they have inbuilt instant mate choice for new wing patterns. Having dissected their DNA, we need to put *Heliconius* back into the wild, to see how these traits play out in the real world. How effective is this mimicry they have been so ingenious in perfecting?

The great dream of biology is to understand nature in the round. Not just to know how genes produce the form and behaviour of creatures, but to see how they function in the ecosystem. That the butterfly world is a microcosm is obvious to anyone who visits a butterfly house. Here they are captive, of course: there are no predators; nothing can get in and nothing can get out. But here are their host plants, where the eggs can be laid; here are the pupae, safely sealed up in a cabinet where every half hour or so another butterfly is emerging; here are the courting butter-flies. Transpose all this to the wild, to the South American habitat, and you have the real butterfly world, in which mimicry is put to the test. Each group of species lives in a microhabitat: the little patch of vegeta-tion that has the right food plants for its caterpillars, the passion vines with their toxins – in a sense, the life's blood of *Heliconius*.

When we can genetically transform one species of *Heliconius* into another, as has been done with the sticklebacks, we shall be able to create novel patterns. Then we can go back to the rain forest, where Bates caught the first jinking glimpse of mimicry on the wing, and, no longer innocent of nature's processes, observe how the patterns shape and are shaped by the butterflies' lives: between the passion-flower food plants and the insectivorous birds, jumping from one mimicry ring to another, evolving their minimalist Matissean masterpieces of painted wing patterns. The circle of genes, body patterns, lifestyle and how they evolved together will be completed. One chapter of the book of life will be fully written for the first time.

CHAPTER 16

A SHIFTING SPECTRUM

These animals also escape detection by a very extraordinary, chameleon-like power of changing their colour. They appear to vary their tints according to the nature of the ground over which they pass: when in deep water, their general shade was brownish purple, but when placed on the land, or in shallow water, this dark tint changed into one of a yellowish green.

Charles Darwin, *The Voyage of the Beagle*

From Bates to Cott, mimicry was mostly observed in nature and described. Theory followed and, as Miriam Rothschild noted, experiment lagged a surprisingly long way behind – around 100 years in most cases. In recent work, some old stories have added new twists and some have experienced reversal. The fact that life is opportunistic, with no set goals other than to survive and reproduce – life is always, in Picasso's words, 'trying new things' – means that many stories of mimicry and camouflage are more complex than was originally thought. Tracing mimicry and camouflage from the time of Bates and Wallace in the 1850s, we have now come up to date. Contemporary research is reopening old casebooks, using modern electronic technology to understand the senses of animals, devising new experiments in the field to test mimicry. The results are often surprising.

When, in 1917, the French Army created the first camouflage unit in a fighting force, they chose as their symbol the best-known example of concealment in nature: the chameleon. In the wider culture, the chameleon is a by-word for colour changes which blend in with the background. In 2006, in Britain, negative political campaigning against the Conservative leader David Cameron featured a cartoon starring Dave-the-Chameleon; this suggested deception and the wearing of false colours. Such broad media campaigns are only possible when an image is widely present in popular consciousness.

But, although most people 'know' that chameleons are well camouflaged and can change colour to suit their background, the experts are less sure. It has been understood for a long time that chameleon colours have some functions unrelated to camouflage. In mating and in territorial combat between males they display their brightest colours, the aim being the opposite of camouflage: maximum display, maximum contrast, maximum brightness.

The deep processes involved in chameleon colour change have only recently been investigated. Animals have visual systems different from ours, often involving a capacity for ultraviolet vision. In chameleons there is one system for brightness and another for colour. The latest research suggests that the dramatic colour changes which have made chameleons famous are only concerned with signalling: in particular, with displaying aggression between males. Background matching for colour does not seem to happen; brightness contrast may be used to some extent for concealment but, strangely, it seems that chameleons have been hiding their flamboyant display behind the illusion that they are trying to be inconspicuous. In their standard coloration, chameleons are reasonably well camouflaged; but when they change colour, contrary to popular myth, they are not making an attempt to blend in better or to match anything in their environment.

The chameleons of the ocean are the octopuses: this is how Darwin described the first octopus he saw. In fact, the colour changes of octopuses surpass those of the chameleon in every respect; and octopuses really do change colour to match their background. We have only in the past ten years discovered that the octopus is a living, breathing, swimming compendium of every camouflage and mimicry technique known. Unlike other camouflaged creatures – which have to make do with the appearance they were born with and seek the best locations and times to maximise their effectiveness and minimise their profile – octopuses and squid can, to a remarkable degree, match almost any environment into which they are placed.

Why *these* animals, and how do they do it? Octopuses and squid are very tempting targets for every large marine predator. They are soft-bodied, have no shell, and an octopus is a large meal. Without their dazzling optical tricks they wouldn't have survived. An octopus's skin is a bit like a living TV screen. Just as a colour TV picture is made up of red, green and blue dots, the octopus has three kinds of colour cell – chromatophores. The colours are black, red and yellow, plus some structural colours, and the cells have shutters so that a colour can be fully or partially displayed or hidden. Not only that, but the skin has warty projections which can be long and spiky, short and rounded or smooth. These give a three-dimensional texture, so that an octopus is equally happy mimicking a barnacle-encrusted stone, a coral, or the smooth skin of a fish.

Aristotle (384–322 BC) – the first naturalist whose work has come down to us – knew and admired the octopus's skill in camouflage. The Mediterranean octopuses were an abundant source of study for him and, in *The History of Animals*, Aristotle made a discovery which was dis-believed until the nineteenth century. In the male, one of the octopus's tentacles is adapted for fertilisation: it is in effect a giant penis, a typical example of nature making use of existing materials.

There is always more to learn about these metamorphic animals. Surprisingly, given their brilliant colour matching, octopuses and squid are colour blind. Roger Hanlon at the Woods Hole Oceanographic Institute, Massachusetts has proved this by showing that, although squid can respond to a black-and-white chequerboard pattern, they cannot see any pattern at all if the squares are yellow and blue. The squid has only one light receptor, and that responds best to green light. This seems odd at first, but it is actu-ally powerful evidence for natural selection. It is the senses of the *predators* that have created the powerful and flexible camouflage systems of the octopus and squid. The octopus controls its patterns by direct nervous control, having first sensed the environment with its acute but colour-blind eyes. The patterns which are produced, triggered by the octopus's brain cells, are the patterns that will fool a predator's nervous system.

An octopus's skin is a marvel of coordinated organs. The chro-matophores are coloured by means of chemical pigments – beneath them come physical structures which produce optical effects, although they themselves have no colour. The first of them are the iridophores. These are stacked multilayers of tough, horn-like material and they produce spectral colours, mostly blue-green, in something like the way a compact disk shows spectral colours. But most interesting of all are the deepest elements: the leucophores. These are reflectors which, in bright light, reflect back whatever colour impinges on them. This is how the octopus compensates for its color-blindness: the leucophores simply reflect back whatever is the ambient light colour.

To add to the colour matching, the octopus senses texture and selects a stippled, mottled or disruptive pattern. Although the octopus can match an enormous number of marine environments with uncanny accuracy, Hanlon believes that the octopus uses only these three modes to match any given environment. Uniform coloration has minimal variation in contrast; mottled coloration has small-scale dark and light patches; while disruptive coloration has large areas of different colours. Finally, the warty skin outgrowths are projected when the background is highly textured or withdrawn if it is smooth.

Octopuses run the gamut of camouflage techniques. First of all, when they are not hungry, they hide in caves. Whilst foraging, they have special strategies such as 'moving rocks'. In this, the legs are held under and the octopus walks on tiptoe, wearing a mottled pattern similar to the surrounding stones and rocks. Besides camouflage, octopuses have startle patterns with fake eyespots, just like the butterflies. And if all else fails, they can emit a black ink cloud – nature's own smokescreen.

The new knowledge about octopuses concerns their ability as mimics. The mimic octopus (*Thaumoctopus mimicus*) was discovered as recently as 1998, off Sulawesi, Indonesia, by photographers rather than biologists. It became a star of TV documentaries* before the scientists arrived – a sign of the times. The mimic octopus appears to imitate a number of fishes and other marine creatures (all toxic) in the local environment. Most of these forms of behaviour remain speculative, but Roger Hanlon has produced convincing evidence of flounder mimicry. When swimming

---

* Video footage of most of the mimicking creatures described in this chapter, including the mimic octopus, can be found on YouTube.

in exposed positions, the mimic octopus holds its tentacles together in one mass, in a very compelling impression of a flounder. Other, less well-attested models, are the lion fish (tentacles splayed in starfish pattern with black and white bands) and the banded sea snake (tentacles again banded but two held in line to create the snake shape). All of these putative mimics are toxic, so this is a case of Batesian mimicry, like that of the swallowtail butterflies – but with the various different mimics occurring not just within the same brood but in the same animal!

The octopus provides strong support for all of the techniques which were so encyclopaedically described by Hugh Cott back in 1940. Perhaps the most controversial of these is disruptive coloration – in his tussles with Thayer, Theodore Roosevelt was most scathing about this aspect. But, apart from blending in with their background, octopuses also use disruptive coloration – they can create Mexican waves of brilliant colour coursing though their bodies. The important thing sometimes is not to look like an octopus at all; and they can produce very bright, highly contrasted patterns to achieve this. They also – Thayer and Cott take a bow – do countershading. Hanlon suggests that the octopus's visual-matching patterns – stippled, mottled or disruptive – are the template for the entire range of camouflage across the animal kingdom.

A fault in early studies of mimicry and camouflage was to take it for granted that the creatures concerned see more or less what we see. The thrill of seeing an orchid flower suddenly resolve itself into an orchid-flower-plus-a-mantis or of a flower spike deconstructing itself into a swarm of insects seemed to brook no argument. But now that we can investigate the visual systems of animals, we see that they are in some respects better than ours, in some respects inferior. Most mammals, for instance, do not have three-colour vision like us. Only our fellow primates, in fact, possess this. Against that, birds and insects often have vision that extends into the ultraviolet, a range of the spectrum we cannot see. How does this affect mimicry and camouflage?

A good test case involves the crab spiders. They always seemed a simple if astonishing case of special resemblance for the purpose of luring their prey: an all-white or all-yellow spider sits in a white or yellow flower and is seen to catch bees which approach the flower for nectar, apparently oblivious to the spider's presence. What could be simpler? But, the more we learn, the more we see that nature doesn't operate on such uncomplicated lines.

The obvious question to ask about bees and crab spiders and flowers is: how does it look to the creatures themselves? We are only spectators, not the protagonists. Lars Chittka at Queen Mary, University of London is a specialist in the visual senses of insects, especially bees. The difference between our visual sense and that of insects and many other creatures, including birds, is that the frequencies they respond to are shifted away from the red towards the blue. In fact they go off the scale. Beyond the darkest blue end of the spectrum (violet) is the ultraviolet: we cannot see this, but the insects can. The flip-side of this is that bees, for instance, cannot see red at all.

Chittka and his colleagues discovered in 2003 that the Australian white crab spider, which lurks in large white daisy flowers, is actually highly *visible* in ultraviolet light. This means that, when a crab spider is on a flower, a bee will see a *brighter* flower than usual. So, far from simply not seeing the spider in the flower, the bee is especially attracted to just those flowers that contain spiders! The 'spider-flowers' are like those wonder whitener detergents in the TV ads: quite unearthly in their brilliance and intended to dazzle, and the bees cannot resist them. At least this is the basic story – it will become more astonishing.

Australia is one of the crab spider's main territories, and Mariella Herberstein at Macquarie University, Sydney has discovered further subtleties in the crab spider story. Insects live in a multi-media wrap-around sensorium and it is not wise to consider one sense in isolation. As Rothschild stressed, odour is a very important sense for insects and it turns out that bees and crab spiders alike are attracted to particular flowers by their odour. An odourless flower will attract neither spiders nor bees. Presumably the bees use odour as an indicator of the fitness of the flower – a bee doesn't want to waste time visiting unhealthy flowers which have little nectar, and odour is the key. The spiders meanwhile have 'deduced' that the best flower to sit in is an odorous one, which will attract the most bees.

But the strangest of Herberstein's discoveries is that Australian native bees, which have co-evolved with the crab spiders, *have worked them out*. Although bees are still attracted to that whiter-than-white glow, they lose interest for some unknown reason when they get closer. As Herberstein says:

In the coevolutionary arms race between crab spider and native pollinator, it appears that the native pollinator currently has the upper
hand and it may be that spiders, in turn, will respond to this predator-
avoidance adaptation by reducing conspicuousness or by exploiting a
different sense modality to attract native prey.

The struggle between the bees and the spiders is one of the arms races
that Cott and Dawkins have described. If one party gains an advantage,
the other has to evolve counter-measures. Deception and counter-
deception in nature and human culture follow the same rules.

So, if evolution is happening here before our eyes and the spiders can
no longer prey on Australian bees, how come they are still here? What
are they eating? Besides the native bees, Australia has a large population
of European bees, and these are still lured by the spiders. Herberstein
comments: 'I think these crab spiders are one of the few examples of
native Australians that benefited from European settlement.'

The 150-year-old phenomenon of Batesian mimicry (harmless
species imitating toxic ones to evade predators) is still producing
surprises. One of the most sinuous cases involves the toxic coral snakes
and their non-toxic mimics, the kingsnakes. Coral snake mimicry is bold,
the snakes being large, as mimetic creatures go, and dramatically banded
in the traditional warning colours of red, yellow and black. The
venomous coral snakes are so well advertised that, wherever they live,
harmless mimics are also found. Surprisingly – and handily for us – the
mimics can match the colours, but not the order in which they appear.
Although there are a few exceptions (be careful!), the following ditty can
be used to distinguish the goodies from the baddies:

Red next to yellow
Kill a fellow.
Red next to black,
Venom lack.

It is strange that, in mimicking the coral snake pattern, nature has not
replicated the same order of the coloured bands. Presumably there is no
creature with vision sufficiently sophisticated to spot the difference.

David Pfennig at the University of North Carolina, who has been collecting coral snakes since he was a kid, comments: 'There may not have been strong selection on mimics to match their model more faithfully, because what they already have seems to work perfectly well.'

Pfennig's research on coral snakes and their kingsnake mimics has given us a new understanding of the complexities of Batesian mimicry, for which they are good model organisms. Coral snakes are so seriously, death-dealingly toxic that predators cannot afford to make a mistake with them. There are puzzles concerning the range of the model venomous coral snakes and their non-toxic kingsnake mimics. For example, the non-venomous scarlet kingsnake is found several hundred kilometres beyond the range of its toxic model, the eastern coral snake. Mimicry is expected to cease where the model is absent, because the mimetic patterns then make an animal highly conspicuous, and conspicuousness coupled with no protection is asking for trouble from predators.

Pfennig found that, in the region where only the non-toxic snake lives, mimicry is indeed breaking down; but how did these snakes get there in the first place? Subtle DNA studies have revealed that males, but not females, have migrated from the area shared by mimics and models.

But now comes another twist. If mimics beyond the range of the model are losing their mimetic patterns, you might expect mimicry to fall off gradually in moving from the territory with plenty of models to the region with none. But Pfennig found that the best mimicry is actually found on the *edge* of the range, where models are few and far between. This has been established experimentally and, after the fact, it is easy to rationalise. Where models are plentiful, perfect mimicry isn't necessary. The models are so dangerous that, when they are common, anything that looks remotely like them is shunned. But on the edge, where models are rare, predators might be inclined to have a go at anything which is less than perfect. After all, not far away, the only snakes that look remotely like this are harmless mimics deprived of their protective model. So here the kingsnakes have to be spot-on.

If the coral snake story is still evolving, the oldest story in natural camouflage – the case of industrial pollution and the peppered moth (*Biston betularia*) – has come full circle. Beginning in the late nineteenth century, this classic tale has had two periods of fame, the first as the prime example of evolution in action and the second as a battleground between biologists and creationists. The first seemed so simple. Many species – the panther, for example – can produce dark forms. A fairly simple genetic mechanism turns most of the cells black, thus obliterating whatever pattern the surface of the creature would normally possess. This is known as melanism. Obviously, lightly patterned and black creatures are perceived very differently against different backgrounds by predators. The peppered moth, as its name implies, is a lightly coloured, subtly patterned moth, well camouflaged against the patterns of tree bark, particularly if it is covered in lichens. When, in the mid-nineteenth century, the trees of industrial areas such as Manchester turned black with pollution, killing the lichens, increasing numbers of melanic peppered moths were found. The moths didn't 'turn black': the black form, which was always naturally produced in small numbers (it is a polymorphism) and which had a low chance of survival, suddenly found its situation transformed. It was now the better camouflaged moth. It started to increase, and the typical form started to decline.

There were some anomalies which slightly spoilt this simple picture. In rural areas, the typical form was expected to continue to be dominant and mostly it was, but in East Anglia, which had no industry worth a mention, the melanic form came to make up over 90 per cent of the population. This was thought to be caused by wind-borne industrial pollution from the Midlands.

But the general pattern was clear: in industrial areas, the melanic form of the moth had dramatically increased, and the most likely explanation was that birds could easily pick light moths off dark trees and vice versa. In 1953, Bernard Kettlewell, a researcher in E. B. Ford's department at Oxford University, had set out to test this hypothesis.

In another of those fateful encounters – Bates and Wallace; Clarke and Sheppard – Kettlewell had come across his lepidopterological destiny in a chance meeting. In 1937, he was a thirty-one-year-old medical doctor from Cranleigh, Surrey, and a passionate amateur lepidopterist. On a butterfly excursion to the Black Wood of Rannoch in Scotland he bumped into E. B. Ford. A correspondence ensued, and fourteen years later Kettlewell came to Ford's department in Oxford to work, thanks to yet another of those grants Ford was so adept at securing from the

Nuffield Foundation. They were an odd couple: Kettlewell an untidy, untheoretical, relatively uneducated field researcher; Ford the pseudo-pontiff of ecological genetics. But Kettlewell had the entomological equivalent of green fingers; he was the practical foil to Ford's genetic theorising. He was there to fulfil part of Ford's grand design.

Kettlewell released the two forms of *Biston betularia* – the black and the speckled – in two different locations: a wood near industrial Birmingham, and one in rural Dorset. He duly found that the melanic form did have an advantage in Birmingham, as did the speckled form in Dorset. It was all very neat, and Kettlewell become something of a celebrity. Ford was as proud of this work as he was of Clarke and Sheppard's.

The story of Kettlewell and the peppered moth is told in blow-by-blow detail, albeit from a relentlessly negative viewpoint, in Judith Hooper's *Of Moths and Men* (2002). The fact is that, although Kettlewell's conclusions were almost certainly correct, the experiments were in many ways unsound. As the most familiar example of camouflage in nature, the peppered moth ought to have been a front-runner for the uncovering of the genetic mechanism behind this ability of the environment to write a pattern (however generalised, as it is in this case) onto a creature. Instead, the peppered moth has been dogged by doubt until very recently.

Following Kettlewell, various other studies of the peppered moth and melanism were made. For a time, it seemed that factors besides predation by birds might have been involved in the population differences. Many moths hardly travel from the place where they are born: the peppered moth goes further than most. Such migration mixes genes. In 1982, a paper in *Nature* cast doubts on predation as the principal cause of the population changes, suggesting that industrial pollution had caused direct changes, perhaps by mutation, in the moths, citing as evidence the fact that many animal species had become melanic in the industrial era, including some, such as cats and pigeons, that had no predators.

Since the Second World War, cleaner air has meant that the peppered moth story has gone into reverse, with the original, pale, typical form now in the ascendant. Year on year, in most locations, the ratio of black moths to speckled declines. This is accepted by all, but the *Nature* paper claimed that the decline of the black moths was more rapid than could be accounted for by the predation hypothesis. The natural form is camou-flaged against lichen, and the peppered moth recovered more rapidly than the lichen it was supposed to be hiding against. The conclusion was that 'non-visual selective forces whose nature is as yet unknown' must be considered and that 'even for a melanic moth life is never as simple as made out in the textbooks'. This was the prime and most simple case of camouflage in nature. Was nothing sacred?

The loose ends in the peppered moth saga were a matter of the normal cut-and-thrust of research until 1998, when, in a strange concate-nation of events, the story flit the confined cage of scientific debate and entered the dangerous viral world of creationist controversy. In that year, the late Michael Majerus from the Department of Genetics at Cambridge University published a book called *Melanism: Evolution in Action*. He reviewed the peppered moth story, and, despite a rather star-tling way of opening the case – 'The findings of these scientists [researchers into the peppered moth after Kettlewell] show that the précised description of the peppered moth story is wrong, inaccurate, or incomplete, with respect to most of the story's components parts' – concluded: 'In my view, the huge wealth of additional data obtained since Kettlewell's initial predation papers does not undermine the basic qualitative deductions from that work.'

When *Melanism* was reviewed in *Nature*, on 5 November 1998, Jerry Coyne, an American biologist, seemed to find the wealth of complexity in the book compelling evidence for the abandonment of the textbook story. By now it was widely recognised that Kettlewell's work was flawed in some respects. His experiments were too artificial: sticking dead moths in quite high numbers on tree trunks during the day did not repli-cate the real situation in the wild. Coyne made much play with this, as if everything depended on Kettlewell's work, despite many other studies.

Coyne probably invited the mayhem which later descended by casting his book review in terms of a loss of faith: 'My own reaction resembles the dismay attending my discovery, at the age of six, that it was my father and not Santa who brought the presents on Christmas Eve.' And this in *Nature*! Coyne concluded that 'for the time being we must discard *Biston* as a well understood example of natural selection in action, although it is clearly a case of evolution'.

That last caveat was lost on a certain faction into whose hands Coyne's review fell. By now creationism was in full swing, and its adherents were looking for any stick with which to beat evolution. If a biologist could write like this in *Nature*, clearly the evolutionary house was in such disorder that a full frontal assault seemed likely to be propitious.

Majerus had written: 'My view of the rise and fall of the melanic peppered moth is that differential bird predation in more or less polluted regions, together with migration, are primarily responsible, almost to the exclusion of other factors.' In Coyne's review this 'primarily' became 'probably', and he compounded the offence by adding: 'I would, however, replace "probably" with "perhaps".' In this way, black can become white in three easy stages of Chinese whispers. Majerus's analysis of complexity had become Coyne's loss of faith in Santa Darwin. The *Sunday Telegraph* Science Correspondent, Robert Matthews, saw Coyne's piece and pushed the story even further in the wrong direction:

> Evolution experts are quietly admitting that one of their most cherished examples of Darwin's theory, the rise and fall of the peppered moth, is based on a series of scientific blunders. Experiments using the moth in the Fifties and long believed to prove the truth of natural selection are now thought to be worthless, having been designed to come up with the 'right' answer.

That slippery declension, from 'primarily responsible, almost to the exclusion of other factors' to 'now thought to be worthless', had launched a storm, like the proverbial butterfly flapping its wings in Florida to start a hurricane in Oregon: the Butterfly Effect. Matthews' article is still fuelling creationist websites, ten years later.

The story lurched further into notoriety in 2002 with the publication of Judith Hooper's *Of Moths and Men*. Hooper is a science writer who turned the peppered moth story and the scientists who studied it into a tale of incompetence verging on fraud. And there was a plucky lone researcher, the American Ted Sergeant, frozen out by a snooty Oxonian establishment – E. B. Ford and his gang of toffs with butterfly nets again. For Hooper, Ford's English school of ecological genetics was a pernicious force, which had obscured the truth of the peppered moth story.

Her attack was *ad hominem*: some moth men had the 'stunted social skills of the more monomaniacal computer hackers, going about with mis-buttoned shirts and uncombed hair'. That doesn't sound much like toffs; and since when did personal untidiness seriously compromise the results of scientific research? Did Einstein bother to comb his hair? Hooper was no creationist, but her book was useful to them.

Majerus, whose book had inadvertently launched a bandwagon now out of control, was deeply disturbed by what had happened. He set out to rectify the situation by devising a seven-year experiment in Madingley Wood near Cambridge, to correct the procedural errors of Kettlewell and to answer some of Hooper's queries. Hooper had interviewed Majerus for her book but was not satisfied with his explanations.

One complaint against Kettlewell which Hooper made much of was that birds were assumed to be the cause of different rates of predation in moths, although this was hard to prove. Birds hunt by day and have visual systems similar to ours. Hooper had suggested that night-flying bats, which hunt by a high-frequency sound echo-location system, might be much more serious predators of moths.* Majerus ran an experiment just to try to satisfy Hooper. In over 200 kills of both forms of the moth, he found no difference in bat-kill rates. So predation by bats falls equally on the dark and light forms of *Biston*. If one form is gaining ground over the other, this is not due to predation by bats.

Another flaw of Kettlewell's made much of in subsequent years was that he stuck his moths onto the *trunks* of trees. Many experts commented that, wherever peppered moths rest during the day, it isn't on tree trunks. So Majerus set out to discover their day-time haunts. He found just over

---

* Some moths have evolved countermeasures against bats. Like modern fighter planes with their radar jamming systems, they have found ways to turn the bats' ultrasound against them. Tiger moths are a case in point. Some of them fly by day and night. They sequester pyrrolizidine alka-loids from *Senecio* and *Crotalaria* plants which render them unpalatable to predators in the classic manner, and they have developed bright, intricate, red and brown patterning. Their caterpillars are also warningly coloured, to be protected by day. But how is it possible to tell a bat which hunts by echolocation that its next mouthful is going to be spat out – so don't even think about it? The night-flying tiger moths can emit bat-like ultrasonic clicks from specialised blisters on the thorax. The sound produced by the moth is by way of educating the bat to the unpalatability of a species which sounds just like that. The moth, in other words, is simply saying: 'I am making a noise too and it is simply to tell you that I taste nasty.' This is the colourful, distasteful species concept translated into sound.

50 per cent of them on lateral branches, but 37 per cent *were* found on tree trunks, it's just that they were quite hard to spot! It is odd that people had been arguing against Kettlewell on these grounds for 50 years yet he was not so very wrong.

But Majerus's critical experiment concerned predation. He marked, released and recaptured the moths: around 21 per cent of the light form and 29 per cent of the dark form were taken by predators. Over the period 2001–7, the proportion of the dark form declined from about 12 to 1 per cent (in the years 1957–64 it had been as high as 94.8 per cent),* and these relative predation rates were what you'd expect them to be to produce such a decline. The peppered moth story, was, it seemed, safe after all.

Majerus presented his findings at the European Society for Evolutionary Biology conference at Uppsala in August 2007. Jerry Coyne told Majerus after the meeting that he 'hadn't meant to say it wasn't a good example of evolution' and that he would 'be very happy now having seen [Majerus's] data to say that actually this is now a well understood example of Darwinian evolution in action and should be back in the biology books'.

Sadly, Mike Majerus's triumphant reinstatement of the peppered moth was to prove a personal swansong. He became ill with mesothelioma in November 2008 and died in January 2009.

In the case of the peppered moth, firstly, the reporting of science spiralled out of control; secondly, the creationist lobby was waiting to pounce on the misapprehension generated. Scientists are actively trying to make sense of highly complex phenomena: they don't claim to be right about everything all the time. There are always imperfections which cry out for further explanation. That is what keeps science on the road. But creationists are only looking for flaws; they are not trying to make a synthesis because they already know the answer – one that is not an answer at all, because it cannot explain anything which is testable experimentally. If one doubt is cast on a piece of science which is a minute part of a vast interlocking system, that is all that creationists want to hear. For

---

* The most thorough record of the decline of the dark industrial form of the peppered moth comes from Sir Cyril Clarke's annual records for 1959–95 at his home on the Wirral peninsula, downwind of Lancashire industry: the decline was similar to that in Cambridge: from over 90 per cent to under 20.

all the biological processes known and understood in evolutionary terms, creationists have no alternative explanation.

Because, in terms of a human lifespan, the process is so slow, evolution is inevitably the hardest of all biological problems. What is remarkable is how much we do know about it, despite the difficulties. The theory of evolution cannot be discredited just because a single researcher made minor methodological errors in a pair of woods in 1953.

EPILOGUE

Mimicry is imitation but it is also an immensely fertile, creative process. As human beings we learn, first of all, by copying. We value originality very highly, but artists know how much originality owes to imitation and to covering your tracks. T. S. Eliot was only laying bare the secrets of the trade when he said that mature poets don't merely imitate: they steal. Or, as Elvis Costello put it, you start out trying to copy something that fascinates you, fail and, in doing so, come up with something new.

In the Prologue I quoted the art historian Sir Ernst Gombrich's delight at finding expressive visual gestures in the natural world. Here, long before anyone painted a deer on a cave wall or carved a bulbous fertility goddess, there were pictorial signs signifying: 'I am dangerous, don't even think about trying to eat me', or 'Look at my spread, brocaded fan tail with its myriad iridescent whirlpool eyespots, doesn't it make you want to . . .'.

Some artists such as the writer Vladimir Nabokov and the painter Abbott Thayer have been highly aware of nature's mimickings and textural facsimiles and have woven them into their work; in Nabokov's case, he wrestled with the deeper meaning of these resemblances. Others, such as Picasso, have blundered through the world, ransacking nature and human detritus for materials, copying natural creation in a more organic way. Picasso was slyly aware of what he was doing, seeing his influence in First World War camouflage, setting himself up as a creator on a par with God. His wanton, punning creativity was perfectly displayed in the sculptures he produced in the early years after the

Second World War. In an echo of the desert *camoufleurs* making their dummy trains and tanks with palm fronds and recycled petrol cans, Picasso scoured the dumps of Vallauris, in the South of France, for the found objects he would transform into animated beings. An old wicker basket forms the ribcage of a goat; another basket is the torso of a little girl skipping; rusted screws make up the feet of a little owl. By far the most wondrous transformation is Picasso's reading of a baboon's face in the bonnet and windscreen of his son's toy car: actually two cars, which, placed back to back, wheel to wheel, create the head. The incongruity of it is reminiscent of the warning 'face' on a *Papilio troilus* caterpillar. And, besides, it is a kindred creature to the desert Cannibals – objects cannibalised so as to make, visually, something completely different. Nature and artists make use of what is to hand and transform it into something rich and strange. They are bricoleurs, tinkerers.

Art has threaded though my story either because it really is a part of it, as in Thayer and Peter Scott, or because it illuminates processes in the natural world. And war brought together three realms in camouflage. As Cott was the first to recognise, the perpetual attrition that is nature and human military conflict have some things in common; hence the tricks of camouflage and disguise which reached a peak in the deceptions of El Alamein in the Second World War.

Some natural creatures also have their own desert campaigns. To the rock pocket mice of the deserts of the southern USA has fallen the honour of being the first of nature's camouflaged animals for which the genetic mechanism is known. Rock pocket mice live in rocky habitats in southern Arizona, New Mexico and northern Mexico. Most of them are sandy coloured with a white countershaded belly, matching the rocks. But the region also has dark volcanic outcrops less than a million years old. The mice on these outcrops are almost universally dark. They are preyed upon by owls which, experiments have shown, can, even at night, differentiate light coloured animals from dark-coloured ones. So at the ecological level the pattern is clear. On these laval outcrops, dark forms of the mice have been selected and their evolution has taken place in less than 1 million years.

In 2003 we learnt that the mice's colouring, from sandy to dark brown, can be flipped by a single gene. In fact, it is possible that the mice from the Pinacates lava flow in Arizona can be flipped by just a single base-pair substitution in a signalling gene. This is the absolute zero of genetic change, the quantum, smaller than which it is not possible to go. Here at last is our smoking gun of adaptive change: a correlation between genetic make-up and a new pattern of camouflage. Darwin would have been thrilled to know that, around the time of his big anniversary (200 years since his birth and 150 years since publication of *On the Origin of Species*), the great intangibles in his scheme of evolution by natural selection are becoming tangible. And this first revealed mechanism is a micro-mutation, as he would have expected; but it managed to produce a huge change in the animal, painting the entire creature in a different colour. Painting a mouse brown instead of sandy is kid's stuff compared to the butterflies' abstract masterpieces, but everything starts small: all of life once consisted of single-celled organisms. Butterfly pattern making will follow.

More than any man who has ever lived, Darwin wanted to know nature in all its detail, whatever that proved to be: to tease out the processes of life at every level. He was a reluctant theorist, waiting for twenty years to publish *On the Origin of Species*, and he went on amassing evidence until he died. In the modern period, studies of camouflage have sometimes seemed to mark time. Cott's 1940 book is still often held up as the last word. And in two world wars biologists and artists struggled under the desperate constraints of war to improvise ways of using camouflage and deception. But since the Second World War, the science of warfare has grown up, just as biology has advanced into the genomics era, in which computers are able to piece together an entire genome from thousands of fragments. Interest in military camouflage at first waned after the Second World War, as it always does in peacetime. Night vision binoculars were developed during the Korean War, making purely visual camouflage inadequate, and weapons technology was developing so fast that camouflage seemed a side-show.

But in the 1960s the Vietnam War brought camouflage back onto the agenda. The jungle environment and the guerrilla skills of the North Vietnamese made it essential. Out of practice in camouflage, the Americans at first borrowed a tiger stripe pattern from the South Vietnamese army for their uniforms while they were resuscitating the lost art of camouflage design. Aircraft camouflage also returned, because it was found that the unpainted planes rotted in the steamy atmosphere. If they had to be painted at all, they might as well be camouflaged.

Perhaps the most recognisable camouflage pattern of all is the US Woodland pattern, originally developed for uniforms by the United States Army Engineer Research and Development Laboratory in 1948 but not used until 1967 in Vietnam. Known as ERDL, this pattern was based on a late Second World War German one. The pattern was tweaked in 1981 and became officially known as 'Woodland'. The style has since become a fashion staple and is probably the image that comes to mind for most people when the word camouflage is mentioned. What was originally intended to conceal has had an afterlife as something highly recognisable on the high street.

The great change in military uniform camouflage came through psychological research into perception conducted by the US Army Office of Institutional Research in the 1970s. Woodland had relatively large blotches, but chaos theory showed how natural textures had what we now call a fractal element. This is the random pixelated nature of the images of natural things, for instance plants and earth and rock.

Fractals are a way of describing how natural forms such as clouds, mountain ranges, coastlines, snowflakes, fern leaves and trees differ from simple geometrical shapes. According to the discoverer of fractals, the mathematician Benoit Mandelbrot, a fractal is 'a rough or fragmented geometric shape that can be split into parts, each of which is (at least approximately) a reduced-size copy of the whole'. You can see this in that most amazing vegetable: Romanesco broccoli, in which every little conical pile is made up of other little conical piles, which are made of yet more, on the Russian doll principle. Such patterns can be generated by surprisingly simple mathematical equations and are easily modelled on computer.

With this in mind, Lieutenant Colonel Timothy R. O'Neill developed what he called Dual Tex, in which the larger patterns were decomposed, 'dithered' into a fine-scale pattern, like 'adding leaves to trees' without removing the tree shape. The result was a macropattern which disrupted the shape of the target, making it hard to recognise, and a micropattern which matched the texture of the background, making it hard to detect.

In the 1990s the Canadians took the lead. They realised that, with a micropattern, the traditional curved shapes of the macropattern were not necessary. The straight-sided shapes they used were easier to print. CADPAT (Canadian Disruptive Pattern) was introduced in 1997 and set the benchmark for the new, digital camouflage. The US Marines then developed a similar system, known as MARPAT (Marine Pattern).

MARPAT was patented in 2004, and the patent document is a fascinating résumé of the whole story of military camouflage, beginning with Abbott Thayer, who is generously acknowledged. 'Camouflage', the authors say, 'is an art in the process of becoming a science.'

But the document spells out definitively the limitations of Thayer's absolute belief in nature as a touchstone:

> Strategies based on natural observations often fall short of military requirements. First, animal coloration is often idiosyncratic and keyed to narrow co-evolution of predator and prey in a specific econiche – that is the zebra's stripes tell us more about the visual system of the lion than about usable principles of military camouflage. Second, organisms are limited in the strategies (patterns) they can 'employ'. The coloration patterns of animals reflect survival probabilities over a long period of time passed on [through] genetic advantage. However, animals do not 'design' their appearance; the process is passive and exploits genetic exploitation of random mutations.

Military camouflage has to be broadly applicable. No camouflage can suit every terrain, so there are three standard versions: woodland, desert and urban. Digital camouflage has now come to the entire US Army in the shape of ACUPAT (Army Combat Uniform Pattern). Essentially, all these digital patterns use the same principles. The dyes used are effective both in visible and in infra-red light.

Just as modern Evo Devo research has resolved the ancient dispute in biology between those who believe that evolution works through the small cumulative variation school and those who plump for large discontinuous leaps, MARPAT resolves the military conundrum of whether camouflage should aim for 'invisibility' or for 'unrecognisability'. The answer lies in the optical nervous system. There are two separate visual systems: one for locating an object and the second for identifying it. The two aspects of digital camouflage are addressed to these two systems: the micropattern makes it difficult to detect an object at all, and the macropattern makes it hard to identify.

Digital camouflage and Evo Devo's probings into the connections between genes and mimetic and camouflaged patterns are the beginning, not the end. Until the US military developed their digital camouflage, there was no systematic effort to study camouflage in the natural world and in warfare as a perceptual problem by using modern instrumentation and computer analysis. Thayer and Scott, and even Cott to some extent, were amateurs in this respect. But there is now an interdisciplinary research effort growing up in which biologists, visual psychologists and computer scientists can test the visual systems of prey and predators both in the wild and in laboratory situations. In February 2009 the Royal Society published a theme issue of its prestigious *Transactions* on the subject. The editors wrote:

> In the last few years, there has been an explosion of camouflage studies. The renewed interest in concealment has partly arisen following a growing body of research into warning coloration and mimicry. . . . It is an exciting time to study camouflage.

The mimic octopus, the new insights into chameleon colour change, the developing story of the crab spiders are all prime subjects for this new work. Thayer and Cott, who are glowingly acknowledged by the new researchers as the founders of this discipline, often had to battle with indifference and disbelief. Their work is now inspiring an array of new exploration, and there is now – what they lacked – a forum in which the findings can be shared and advanced.

Camouflage and mimicry belong to the great empire of signs, non-verbal means of communication. As long as there are living things on the earth, they will be finding new ways to hide, to feign threat, to startle, and simply to copy that rather snazzy pattern someone else has fabricated.

NOTES

Abbreviations

AHT1 Smithsonian Archives of American Art: Abbott Handerson Thayer and Thayer family papers, 1881–1950. Series subseries 2.1: Series 2. Abbott Handerson Thayer Correspondence. War Office, London, 1916–17 (Box 1, Folder 70). www.aaa.si.edu/collectionsonline/thayabbo/ (accessed 8 March 2009).

AHT2 Smithsonian Archives of American Art: Abbott Handerson Thayer and Thayer family papers, 1881–1950. Series subseries 2.1: Series 2. Abbott Handerson Thayer Correspondence. Roosevelt, Franklin D., 1917–18 (Box 1, Folder 57). www.aaa.si.edu/collectionsonline/thayabbo/ (accessed 8 March 2009).

DCP Darwin Correspondence Project www.darwinproject.ac.uk (accessed 8 March 2009).

EBF Bodleian Library, Department of Western Manuscripts, Papers of E. B. Ford, MSS. Eng. c. 2656–61.

IWM Imperial War Museum Archives, London.

JGKA Sir John Graham Kerr Archive, Glasgow University.

NA National Archives, Kew, London.

PS Sir Peter Scott papers, Cambridge University Library.

Prologue

p. 1. 'The art historian Sir Ernst Gombrich (1909–2001) was interested in more than paintings': Gombrich (1982), pp. 24–7.

p. 3. 'Camouflage research has for a significant length of time': Martin Stevens and Sami Merilaita, 'Animal camouflage: Current issues and new perspectives', *Philosophical Transactions of the Royal Society B*, 2009, 364, pp. 423–7.

Chapter 1 Darwinians, Mockers and Mimics

p. 6. 'Some of these resemblances are perfectly staggering': Letter from Bates to Darwin, 28 March 1861, in Stecher (1969), p. 14; DCP 3104.

p. 6. 'Mr Bates had come a long way': Bates (1863).

p. 8. 'Amongst the lower trees and bushes': ibid., Vol. 1, pp. 247–8.

p. 8. 'He was a prime example of the largely self-educated Victorian': in Stecher (1969), pp. 2–5; Moon (1976); E. Clodd, 'Memoir', in Bates (1892), pp. xvii–lxxxix.

p. 9. 'as the journal of a scientific traveller': Wallace (1908), p. 144.

p. 9.   'At this period too, vast numbers of trees add their tribute of beauty': Edwards (1847), p. 45.

p. 10.   'When the festival happens on moonlit nights': Bates (1863), Vol. 1, pp. 90–1.

p. 11.   'The hot moist mouldy air': ibid., p. 6.

p. 12.   'In the shady ravines of the forest': H. W. Bates, 'Excursion to St Paulo', 5 September 1858, *The Zoologist*, 1858, xvi, pp. 6165–6.

p. 13.   'I am as fond of Latin': Bates (1892), p. xviii.

p. 13.   'I now had a dull time of it in Barra': Wallace (1853), pp. 173–5.

p. 14.   'It now presented a magnificent and awful sight': Wallace (1908), p. 155.

p. 15.   'Unlike Bates and Wallace, Darwin': Desmond and Moore (1992). Except where otherwise referenced, biographical information on Darwin comes from this source.

p. 15.   'shall we conjecture': Darwin (1796), p. 556.

p. 15.   'some dissipated low-minded young men': Darwin (2002), p. 31.

p. 15.   'You care for nothing': ibid., p. 10.

p. 16.   'One day, on tearing off some old bark': ibid., p. 33.

p. 16.   'from not being able to draw': ibid., p. 43.

p. 17.   'Looking backwards, I can now perceive': ibid.

p. 17.   'undermining and hawling down the largest member': Owen (1994), p. 106.

p. 18.   'It has been mentioned': Darwin (1989), p. 287.

p. 18.   'A struggle for existence': Darwin (1985), pp. 116–17.

p. 19.   'When we look at the plants and bushes': ibid., pp. 125–6.

p. 20.   'Ask, as I have asked': ibid., p. 88.

p. 20.   'How fleeting are the wishes': ibid., p. 133.

p. 22.   'Meanwhile Wallace was travelling again': Wallace (1908), pp. 175–202; Wallace (1869).

p. 23.   'Your words have come true': letter from Darwin to Lyell, 18 June 1858: DCP 2285.

p. 23.   'I would far rather burn my whole book': letter from Darwin to Lyell, 25 June 1858: DCP 2294.

p. 23.   'been marked by any of those striking discoveries': Browne (2002), p. 42.

p. 24.   'The paper was published': C. R. Darwin and A. R. Wallace, 1858. 'On the tendency of species to form varieties; and on the perpetuation of varieties and species by natural means of selection' [Read 1 July], *Journal of the Proceedings of the Linnean Society of London. Zoology*, 1858, 3, pp. 46–50.

p. 24.   'If the question is put to me': Desmond and Moore (1992), p. 497.

p. 25.   'As to ordinary Entomologists': letter from Bates to Darwin, 24 November 1862: in Stecher (1969), pp. 37–8; DCP 3825.

p. 25.   'It is not difficult to divine the meaning': Henry Walter Bates, 'Contributions to an insect fauna of the Amazons Valley', *Transactions of the Linnean Society*, 1862, 23, pp. 507–8.

p. 27.   'It is clear, therefore': ibid., p. 512.

p. 27.   'one of the most perplexing problems': letter from Darwin to Bates, 3 December 1861: in Stecher (1969), p. 19; DCP 3338.

p. 27.   'We can understand resemblances': Charles Darwin, 'A review of H. W. Bates' paper on "mimetic butterflies" ', in Darwin (1977), pp. 87–92.

p. 28.   'here we have an excellent illustration of natural selection': Darwin (1951), p. 492.

p. 28.   'I believe therefore': Henry Walter Bates, 'Contributions to an insect fauna of the Amazons Valley', *Transactions of the Linnean Society*, 1862, 23, p. 508.

p. 28.   'They seem far richer in facts': letter from Darwin to Bates, 26 March 1861: DCP 3100.

p. 28.   'I am quite convinced that insects offer better and clearer illustration': letter from Bates to Darwin, 28 March 1861: in Stecher (1969), p. 11; DCP 3104.

p. 29.   'It may be said, therefore, that on these expanded membranes': Bates (1863), Vol. 2, p. 346.

p. 29.   'It sometimes happens, as in the present instance': ibid., Vol. 1, p. 261.

p. 30.   'Whilst reading and reflecting': Charles Darwin, 'A review of H. W. Bates' paper on "mimetic butterflies" ', in Darwin (1977), p. 92.

p. 30. 'It is really curiously satisfactory': letter from Darwin to Hooker, 26 March 1862: DCP 3483.

p. 30. 'It is most curious': letter from Bates to Darwin, 2 May 1863: in Stecher (1969), p. 44; DCP 4138.

Chapter 2 Swallowtails and Amazon

p. 31. 'Suppose that a blue-eyed, flaxen-haired Saxon man': Alfred Russel Wallace, 'On the phenomena of variation and geographical distribution as illustrated by the Papilionidae of the Malayan region', *Transactions of the Linnean Society of London*, 1865, 25, pp. 10–11 (ft).

p. 32. 'Wallace saw male and female swallowtails': ibid., p. 6.

p. 32. 'Trimen presented his findings': Roland Trimen, 'On some remarkable Mimetic Analogies among African Butterflies', ibid., 1869, 26, Pt 3, pp. 497–522.

p. 33. 'it would require a stretch of the imagination': Roland Trimen, 'Observations on the Case of *Papilio Merope*', *Transactions of the Entomological Society of London*, 1874, Pt 1, p. 140.

p. 33. 'Her mate is not generally so early': J. P. Mansel Weale, 'Notes on the Habits of *Papilio Merope*', ibid., pp. 131–2.

p. 34. 'It is with reluctance': Roland Trimen, 'Observations on the Case of *Papilio Merope*', ibid., p. 141.

p. 35. 'The final proof had to wait': Roland Trimen, 'Notes on the Capture of the Paired Sexes of *Papilio Cenea*', ibid., 1881, pp. 169–70.

p. 36. 'I know not how, or to whom': letter from Wallace to Bates, 24 December 1860: in Wallace (1908), p. 197.

p. 36. 'Why are caterpillars sometimes so beautifully and artistically coloured': Wallace (1908), p. 227.

p. 37. 'Now, as a *white* moth is as conspicuous': ibid., p. 228.

p. 37. 'My dear Wallace, – Bates was quite right': letter from Darwin to Wallace, 26 February 1867: in Wallace (1908), p. 229.

p. 38. 'A horse is trained to certain paces': Darwin (1868), p. 445.

p. 38. 'I, though agreeing with him': Wallace (1908), p. 236.

p. 39. 'similar adaptation of all': A. R. Wallace, 'Dr Fritz Müller on some difficult cases of mimicry', *Nature*, 1882, 26, p. 86.

p. 39. 'Fritz Müller (1821–97)': West (2003).

p. 40. 'not improbable that a butterfly endowed': letter from Dr Fritz Müller to Herman Müller, 29 March 1870, in West (2003), p. 224.

p. 40. 'The same view perhaps applies in part to gaudy butterflies': letter from Darwin to Müller, 28 August 1870: British Library, London, BL Loan 10, 33.

p. 41. 'It is remarkable how one racks one's brains': West (2003), p. 235.

p. 42. 'In 1879, Müller's new form of mimicry': F. Müller, '*Ituna* and *Thyridia*; A remarkable case of mimicry in butterflies', *Transactions of the Entomological Society of London*, 1879, pp. xx–xxix (translated by Ralph Meldola from the original German article in *Kosmos*, May 1879, p. 100).

p. 42. 'Wallace came to realise the force': A. R. Wallace, 'Dr Fritz Müller on some difficult cases of mimicry', *Nature*, 25 May 1882, 26, pp. 86–7.

p. 42. 'I have long looked on him as the best observer in the world': letter from Darwin to E. L. Krause, 28 November 1880: DCP 12871.

Chapter 3 Delight in Deception

p. 43. 'Every naturalist traveller appears to have some instance to relate': Beddard (1892), pp. 109–10.

p. 43. 'In March 1878, on the island of Java': Forbes (1989), pp. 63–5.

p. 44. 'How strange it is': ibid., p. 216.

p. 44. 'There is no doubt that the spider must have acquired this mimicking habit': ibid.

p. 45. 'It seems to me, on the contrary': O. Pickard-Cambridge, 'On two new genera of spiders', *Proceedings of the Zoological Society of London*, 1884, Pt 2, p. 197.

p. 45. 'expressly disclaimed the idea': 'the similitude is so exact'; 'Is not its exactness': ibid., p. 199.

p. 46. 'The trade of chemist': Levi (1985), pp. 180–1.

p. 46. 'On picking up from the stony ground': Burchell (1822), p. 310.

p. 46. 'a very characteristic form': *Gardeners' Chronicle*, 24 February 1900, pp. 113–15.

p. 47. '*Pêche de Folha* leaf fishes of Brazil': Cott (1940), pp. 311–13.

p. 47. 'The stone fishes, which use their brilliant camouflage to attack': Sutherland (1983), pp. 400–10.

p. 47. 'The leafy sea dragon being the finest': Cott (1940), pp. 341–2.

p. 47. 'Wallace had drawn attention to the leaf butterfly, *Kallima*': Wallace (1869), Vol. 1, p. 130.

p. 47. '[It] disappears like magic': Darwin (1871), p. 392.

p. 47. 'a much less perfect imitation would be ample': Lull (1917), p. 246.

p. 47. 'Gregory was walking in woods near the Kibwezi River': Gregory (1896), pp. 273–5.

p. 49. 'In 1862, working on the pollination of orchids': Darwin (1862), pp. 197–203.

p. 49. 'No such moth was known': L. W. Rothschild and K. Jordan, 'A revision of the lepidopterous family Sphingidae', *Novitates Zoologicae*, 1903, 9 (Suppl.), pp. 1–972.

p. 49. 'I was attracted to a bush of the "Straits Rhododendron" ': N. Annandale, 'Observations on the habits and natural surroundings of insects made during the "Skeat Expedition" to the Malay Peninsula, 1899–1900', *Proceedings of the Zoological Society of London*, 1900, pp. 839–48.

p. 50. 'aroused a lifelong delight in the facts and theories of Protective Resemblance': G. D. Hale-Carpenter, 'Edward Bagnall Poulton, 1856–1943', *Obituary Notices of Fellows of the Royal Society*, 1944, 4, pp. 655–80.

p. 50. 'Said the Scientist to the Protoplasm': British Association, York, 5 September 1932. Papers of Sir Edward Bagnall Poulton, Bodleian Library, GB 161. Notebook 2.

p. 50. 'he could forgive anything': G. D. Hale-Carpenter, 'Edward Bagnall Poulton, 1865–1943', *Obituary Notices of Fellows of the Royal Society*, 1944, 4, p. 658.

p. 51. 'A Batesian mimic may be compared to an unscrupulous tradesman': Poulton, quoted in Hardy (1965), p. 145.

p. 51. 'protective resemblance – special and general': Poulton (1890), pp. 24–283.

p. 53. 'The defenceless character of the group as a whole': ibid., pp. 264–5.

p. 53. 'natural selection not only tends to pick out and preserve the forms that have protective resemblances': Belt (1874), p. 383.

p. 54. 'For Poulton a species was': Edward Bagnall Poulton, 'What is a Species?', *Proceedings of the Entomological Society of London*, 1903; pp. lxxvii–cxvi.

p. 54. 'The only distinction between species': Darwin (1985), p. 455.

p. 54. 'as it were, trembling on the edge': Poulton (1908), p. 76.

p. 55. 'A *causus belli* was the caterpillar of the lobster moth': Poulton (1890), pp. 278–82.

p. 55. 'Thus the larva of *Stauropus* is supposed to mimic': W. L. McAtee, 'The experimental method of testing the efficiency of warning and cryptic coloration in protecting animals from their enemies', *Proceedings of the Academy of Natural Sciences of Philadelphia*, 1912, 64, pp. 302–3.

Chapter 4 Pangenesis

p. 57. 'To solve the problem of the forms of living things': Bateson (1894), p. 1.

p. 58. 'Darwin was asking his readers': Carroll (2008), pp. 31–2.

p. 58. 'it is called the cell': John S. Karling, 'Schleiden's contribution to cell theory', *American Naturalist*, 1939, 73, pp. 517–37.

p. 58. 'he claimed that every living cell': Robert P. Wagner, 'Rudolph Virchow and the genetic basis of somatic ecology', *Genetics*, 1999, pp. 151, 917–20.

p. 58.   'He wrestled with these ideas': Darwin (1868).

p. 59.   'In 1878, he saw spindly structures': Carlson (2004), pp. 24–6.

p. 59.   'Mendel showed that there were some simple rules': Henig (2000), pp. 67–89.

p. 61.   'There is a persistent legend': Andrew Sclater, 'The extent of Darwin's knowledge of Mendel', *Georgia Journal of Science*, 2003, 61 (no. 3), pp. 134–7.

p. 62.   'new forms of *Oenothera* appeared quite often': de Vries (1909), pp. 217–59.

p. 63.   'William Bateson did not believe that evolution': Bateson (1894), pp. v–x.

p. 64.   'Bateson's ideas entered the field of mimicry': F. A. E. Crew, 'Reginald Crundall Punnett', *Biographical Memoirs of Fellows of the Royal Society*, 1967, 13, pp. 309–26.

p. 64.   'In 1905/6, Bateson and Punnett discovered linkage': W. Bateson, R. Saunders and R. C. Punnett, 'Further experiments in inheritance in sweet peas and stocks', *Proceedings of the Royal Society B*, 1906, 77, pp. 236–8.

p. 64.   'The American Thomas Hunt Morgan': Allen (1978), pp. 144–64.

p. 66.   'A definite small change in the composition of the pigment': Punnett (1915), p. 151.

p. 67.   'a structure of caverns and ribs is revealed': Helen Ghiradella, 'Hairs, bristles, and scales', in Harrison and Locke (eds) (1999), pp. 257–87.

p. 67.   'On this view natural selection is a real factor': Punnett (1915), p. 152.

p. 68.   'He cited as evidence a new dwarf sweet pea variety': ibid., p. 91.

p. 68.   'I have always recognised that the first variation': E. B. Poulton, 'Mimicry and the inheritance of small variations': *Bedrock*, 1913, 2 (no. 3), p. 301.

p. 69.   'Here he began the serious process': Punnett (1915), pp. 146–51.

p. 69.   'H. G. Wells might have had Poulton and Punnett in mind': Wells (1998), pp. 84–91.

p. 70.   'For twenty years he had worked': ibid., p. 86.

p. 70.   'Once he saw it quite distinctly': ibid., p. 89.

Chapter 5   On the Wings of Angels

p. 71.   'Only an artist, perhaps, can rightly appreciate': Thayer (1909), p. 3.

p. 71.   'Thayer was in some respects a typical New Englander': White (1951).

p. 72.   'the noble woman': Anderson (1982), pp. 73–4.

p. 72.   'Doubtless my lifelong passion for birds': Anderson (1982), p. 60.

p. 72.   'Thayer was much like Thoreau': White (1951), p. 97.

p. 73.   'Animals are painted by Nature': Thayer (1909), p. 14.

p. 73.   'The artist, by skilful use of light and shade': Cott (1940), p. 36.

p. 74.   'In 1896, he published his findings': Abbott H. Thayer, 'The law which underlies protective coloration', *The Auk*, 1896, 13, pp. 124–9.

p. 74.   'Thayer's ideas on countershading': E. B. Poulton, 'The meaning of the white under sides of animals', *Nature*, 1902, 65, pp. 596–7.

p. 74.   'I have set up a pair of models of birds': White (1951), p. 87.

p. 74.   'God as a professional colleague': Anderson (1982), p. 117.

p. 75.   'Our book presents not theories': Thayer (1909), p. 3.

p. 75.   'An *artist* is of course the judge': Abbott H. Thayer, *Popular Science Monthly*, Vol. 79, July–Dec. 1911, p. 26.

p. 75.   'The entire matter has been in the hands of the wrong custodians': Thayer (1909), p. 3.

p. 75.   'The peacock's splendour': ibid., Plate 1.

p. 76.   'These traditionally "showy" birds': ibid., p. 154.

p. 76.   'watching more than a fact': Cortissoz (1923), p. 39.

p. 76.   'fundamentally and very potently obliterative': Thayer (1909), p. 205.

p. 77.   'All patterns and colors whatsoever': Thayer (1909), pp. 5, 16.

p. 77.   'Animals' markings doubtless serve': Thayer (1909), p. 6 (footnote).

p. 77.   'The cases of out-and-out butterfly mimicry': ibid., p. 215.

p. 77.   'Probably this *Heliconius* finds': ibid., p. 214.

p. 77.   'My father's special mission was tasting butterflies': White (1951), p. 108.

p. 78.   'our principal and quite unforgetable exhibit': Cortissoz (1923), p. 37.

p. 78.	'Thayer identified a second mechanism': Thayer (1909), pp. 77–8.
p. 79.	'bold massed patterns': ibid., p. 78.
p. 79.	'Skunks . . . long believed by naturalists'; 'attested beyond question': ibid., p. 148.
p. 79.	'seen against the sky': Thayer (1909), caption to Plate 63, following p. 78.
p. 80.	'He devoted a twenty-page appendix': Roosevelt (1910), pp. 501–20.
p. 80.	'The theory is certainly pushed to preposterous extremes': ibid., p. 502.
p. 80.	'No colour scheme whatever is of much avail': ibid., p. 507.
p. 80.	'Ten steps farther back': ibid., p. 504.
p. 81.	'It is the nearest to 100 per cent error': Abbott H. Thayer, *Popular Science Monthly*, Vol. 79, (July–Dec.) 1911, p. 26.
p. 81.	'In passing I wish to bear testimony': Roosevelt (1909), p. 502 (footnote).
p. 81.	'the famous Armory Show of 1913': Brown (1963).
p. 83.	'an explosion in a shingle factory': ibid., p. 110.
p. 83.	'Thayer and his friends were appalled': ibid., p. 131.
p. 83 fn.	'Nothing but the gross puerility': Blunt (1919–20), p. 743.
p. 84.	'an increasing lateral swing': Thayer papers, in Anderson (1982), p. 33.

Chapter 6 Dazzle in the Dock: The First World War

p. 85.	'For England to see these facts': *New York Tribune*, 13 August 1916; White (1951), pp. 253–8.
p. 86.	'Thayer took out a patent': 'Improvements in process of treating ships and other objects to render them less visible'. UK Patent GB190217989 (A), 1902.
p. 86.	'Thayer believed that white was the key': *United States Hydrographic Bulletin*, 1205, October 1912.
p. 86.	'the extent to which man in his war camouflage falls short': John Graham Kerr, 'Camouflage in Warfare' (review), *Nature*, 1941, 147, p. 759.
p. 86.	'Kerr sent Churchill a three-page outline': copy letter from Kerr to Churchill, 24 September 1914: JGKA/DC6/246.
p. 87.	'communicated to the fleet – confidentially in a general order': copy letter from Kerr to Balfour, 28 June 1915: JGKA/DC6/254.
p. 87.	'has aroused great interest': letter to Kerr, 1 December 1914: JGKA/DC6/249.
p. 87.	'I am doubtful from what I have seen': letter from Kerr to G. T. Beilby, Admiralty, July 1915: JGKA/DC6/255.
p. 88.	'But in July 1915 the Admiralty informed him': copy letter from W. W. Baddeley, Admiralty, to Kerr: JGKA/DC6/256.
p. 88.	'what I have written above are mere statements of scientific fact': copy letter from Kerr to W. W. Baddeley, Admiralty, 18 July 1915: JGKA/DC6/257.
p. 88.	'In June 1916 Kerr wrote to Churchill again': JGKA/DC6/260.
p. 88.	'He then turned his attention to aircraft': copy letter from Kerr to Lloyd George, 28 September 1916: JGKA/DC6/261.
p. 88.	'Thayer was even more agitated than Kerr': White (1951), pp. 132–9.
p. 88.	'The strangest of Thayer's ideas': letter from Thayer to the Secretary of the Admiralty: NA/ADM 1/8412/50.
p. 89.	'Several of these freak methods': Admiralty memo of 2 March 1915 from Captain Thomas Crease: NA/ADM 1/8412/50.
p. 89.	'partly lost his grip': White (1951), p. 161.
p. 89.	'total heavenly triumph': ibid., p. 162.
p. 89.	'I shall be curious to know': letter from John Singer Sargent to Thayer, 31 January 1916: ibid., p. 163.
p. 90.	'The First Lord thinks that possibly': letter from Balfour to Thayer, 23 March 1916: AHT1, fos 3–4.
p. 90.	'In August 1916': *New York Tribune*, 13 August 1916; White (1951), pp. 253–8.
p. 90.	'An artist now enters the story': Wilkinson, (1969), pp. 78–100.

p. 90. 'absolutely a spontaneous one': Committee of Enquiry on Dazzle Painting, 5 November 1919, NA/ADM 245/4.
p. 91. 'I had heard of nothing': ibid.
p. 91. 'The whole aim was to use perspective': ibid.
p. 92. 'proposal by A. H. Thayer': Wilkinson (1969), p. 84.
p. 92. 'He had a very skilled modeller': Committee of Enquiry on Dazzle Painting, 5 November 1919: NA/ADM 254/54.
p. 93. 'When I got over there': ibid.
p. 93. 'In cloudy weather': *New York Tribune*, 13 August 1916; White (1951), pp. 253–8.
p. 94. 'On 7 July 1917 he wrote to Roosevelt': letter from Thayer to Roosevelt: AHT2/ ff. 1–6.
p. 94. 'Roosevelt replied a week later': AHT2/f. 7.
p. 94. 'In April 1918 Thayer tried a different ploy': letter from Roosevelt to Thayer, 2 April 1918: AHT2/f. 13.
p. 94. 'How the hell:' Wilkinson (1969), p. 92.
p. 94. 'The American artist Everett Warner': Everett Warner, 'The science of marine camouflage design', *Transactions of the Illuminating Engineering Society*, 1919, Vol. 14 (no. 5), pp. 215–24.
p. 95. 'On 21 July 1918 the British Commander in Chief': 'Dazzle Painting of Merchant Ships', NA/ADM 1/8533/215.
p. 95. 'while it might have been somewhat visible': Wilkinson (1969), p. 93.
p. 96. 'No definite case on material grounds': 'Dazzle Painting of Merchant Ships', NA/ADM 18533/215.
p. 97. 'The results of ship camouflage in the USA': Lt Harold Van Buskirk, 'Camouflage', *Transactions of the Illuminating Engineering Society*, 1919, 14 (no. 5), p. 229.
p. 97. 'After the war, an Admiralty enquiry': Committee of Enquiry on Dazzle Painting, 27 October 1919, NA/ ADM 245/4.
p. 97. 'the talk Wilkinson gave after the War at the North East Coast Institution of Ship Builders': Lieutenant Commander Norman Wilkinson, 'The dazzle painting of ships', read before the North East Coast Institute of Engineers and Shipbuilders, 10 July 1919: NA/ADM 245/4.
p. 97. 'of course it has a relation': Committee of Enquiry on Dazzle Painting, 27 November 1919: NA/ADM 245/4.
p. 98. 'stultifying the German range finders': letter from Kerr to Churchill, 25 June 1915: JGKA/DC6/253.
p. 98. 'obliterating those sharp details': letter from Kerr to Balfour, 28 June 1915, JGKA/DC6/254.
p. 98. 'For a time in the early stages': Lieutenant Commander Norman Wilkinson, 'The Dazzle Painting of Ships', read before the North East Coast Institute of Engineers and Shipbuilders, 10 July 1919. NA/ADM 245/4.
p. 98. 'I was thinking entirely': Committee of Enquiry on Dazzle Painting, 27 November 1919: NA/ADM 245/4.
p. 98. 'that was called the Thayer system'; 'Do you suggest'; 'the whole of this idea'; 'I make no claim'; 'no where in your letters'; 'Incidental resemblance is no ground': ibid.
p. 99. 'Kerr was informed by the Admiralty': letter from W. W. Baddeley, Admiralty, 2 October 1920: JGKA/DC6/277.
p. 99. 'The records of the Royal Commission': JGKA/DC6/636.
p. 99. 'newspaper paragraphs which date the discovery': *The Times*, 19 May 1919.
p. 99. 'Kerr's letter received a devastating riposte': *The Times*, 9 June 1919.
p. 100. 'the Dazzle Ball, held at the Chelsea Arts Club': *Illustrated London News*, 22 March 1919, 154, pp. 414–15.

Chapter 7 Camouflage and Cubism in the First World War

p. 101. 'In order to deform totally the aspect of the object': Mare (1996), p. 131.
p. 101. 'we should note that the successful camouflage officers': Hardy (1965), p. 129.

p. 102. 'He had sent some samples ahead of him': White (1951), pp. 158–63.

p. 102. 'Although they admitted that your observations': ibid., p. 160.

p. 103. 'I had many suits made up to your pattern': letter from J. Stephens, Director of Equipment and Ordnance Stores, to Thayer, 14 August 1916: AHT1/f. 10.

p. 103. 'I am sending you this': undated letter from Viscount Bryce to Thayer: AHT1/ f. 23.

p. 103. 'It is true that, from 1916 on, snipers': Newark (2007), pp. 81–7.

p. 103. 'Gertrude Stein tells the story': Stein (1948), p. 11.

p. 104. 'I was very happy when, in 1914': Liberman (1960), p. 41.

p. 104. 'familiar with the world of painters': Mare (1996), p. 46.

p. 104. 'the point they [cubism and camouflage] had in common': Penrose (1981a), p. 199.

p. 105. 'Later on in the war, Picasso remarked to Cocteau': ibid.

p. 105. 'by concealing a gun emplacement': Addison (1926), p. 107.

p. 105. 'André Mare was one of the artists': Mare (1996), pp. 46–54, 130–1.

p. 106. 'the painter Solomon J. Solomon': Phillips (1933), pp. 116–213.

p. 106. 'the artists were rated': Addison (1926), p. 108.

p. 106. 'the whole of Flanders': Phillips (1933), p. 136.

p. 106. 'nothing but an effective screen': ibid., p. 132.

p. 107. 'the derelict factory': ibid., p. 133.

p. 107. 'It could be stretched over guns': ibid., p. 139.

p. 107. 'real bark sewn on canvas': Addison (1926), p. 122.

p. 108. 'We must win this war': Phillips (1933), p. 163.

p. 108. 'the various attitudes that might be assumed': ibid., p. 126.

p. 108. 'look twice before calling on his artillery': ibid., p. 127.

p. 108. 'all smashed and rendered useless': ibid., p. 127.

p. 108. 'Back in England, during 1917 and 1918': Solomon (1920).

p. 109. 'they were the very first things': Phillips (1933), p. 180.

p. 109. 'the mural painter Barry Faulkner': the subsequent quotations come from Faulkner (1973), p. 18 and pp. 85–111.

p. 111. 'Thayer's war has a postscript': Thayer (1918), p. vii.

p. 111. 'My strenuous and essential life': letter to William Dutcher, 29 November, 1910: Thayer papers D200, Fr 1153–5.

p. 112. 'The word *Camouflage*': Addison (1926), p. 112.

Chapter 8 Hopeful Monsters?

p. 113. 'There are, on your small wings': 'The Butterfly', Brodsky (1980), p. 69.

p. 114. 'It was known as the "Nymphalid groundplan" ': B. N. Schwanwitsch, 'On the ground-plan of wing-pattern in Nymphalids and certain other families of Rhopalocerous Lepidoptera', *Proceedings of the Zoological Society of London*, ser. B, 1924, 34, pp. 509–28; F. Suffert, 'Zur vergleichende Analyse der Schmetterlingszeichnung', *Biologisches Zentralblatt*, 1927, 47, pp. 385–413.

p. 115. 'owing to the war and revolution': B. N. Schwanwitsch, 'On the groundplan of wing-pattern in nymphalids and certain other families of rhopalocerous butterflies', *Proceedings of the Zoological Society of London*, ser. B, 1924, 34, p. 510.

p. 116. 'But in 1927 Suffert showed that by suppressing some elements': F. Suffert, 'Zur vergleichende Analyse der Schmetterlingszeichnung', *Biologisches Zentralblatt*, 1927, 47, pp. 406–7.

p. 118. 'A man over twenty usually does not know': Goldschmidt (1960), p. 251.

p. 118. 'Goldschmidt echoed Bates': Richard Goldschmidt, 'Mimetic Polymorphisms, a controversial chapter of Darwinism', *Quarterly Review of Biology*, 1945, 20, p. 147.

p. 119. 'Heat-shock and cold-shock applied to pupae': Nijhout (1991), pp. 119–31.

p. 119 fn. 'Just to confirm that eyespots': Adrian Vallin et al., 'Prey survival by predator intimidation: An experimental study of peacock butterfly defence against blue tits', *Proceedings of the Royal Society B*, 2005, 272, pp. 1203–7.

p. 120.   'Goldschmidt reported that the European small tortoise-shell butterfly': Goldschmidt (1938), p. 4.

p. 120.   'For Goldschmidt, these transformations were powerful support': Richard Goldschmidt, 'Mimetic polymorphisms, a controversial chapter of Darwinism', *Quarterly Review of Biology*, 1945, 20, pp. 147–64; 205–30.

p. 121.   'a virulent new form of Lamarckism': Lysenko (1954), pp. 475–6.

p. 122.   'We cannot wait for favours': ibid., p. 476.

p. 122.   'I spoke half jokingly of the hopeful monster': Goldschmidt (1960), p. 318.

p. 123.   'Goldschmidt admitted that he "certainly had struck a hornet's nest"': ibid., p. 324.

p. 124.   'Fisher's book has a chapter on mimicry': Fisher (1930), pp. 146–69.

p. 124.   'Goldschmidt's argument': John Turner, 'The hypothesis that explains mimetic resemblance explains evolution, the gradualist–saltationist schism', in Grene (1983), p. 142.

p. 125.   'consider the cyclopic sheep of Utah': Richard F. Keeler and Wayne Binns, 'Teratogenic components of *Veratrum californicum* (Durand). V', *Teratology*, 1968, 1, pp. 5–10.

p. 126.   'in the evolution of snakes, numbers of vertebrae change': Dawkins (1988), p. 236 (my italics).

p. 126.   'Big Brother, the tyrant': Stephen Jay Gould, 'The return of hopeful monsters', in Gould (1982), p. 186.

p. 126.   'In my own, strongly biased opinion': ibid., p. 192.

p. 126.   'reshuffling or scrambling of the intimate chromosomal architecture': Richard Goldschmidt, 'Evolution, as viewed by one geneticist', *American Scientist*, 1952, 40, pp. 96–7.

p. 126 fn. 'Evidence began to emerge in the 1940s': McClintock (1987).

Chapter 9   The Natural History of the Visual Pun

p. 127.   'If all of these interpretations': Mabey (1983), p. 128.

p. 127.   'From the age of seven': Nabokov (2000), p. 94.

p. 128.   'He variously warmed and cooled': Nabokov (2001), pp. 104–5.

p. 128.   'He [the protagonist's father] told me about the incredible artistic wit': ibid., p. 105.

p. 129.   'a fully qualified, clearly talented, duly employed professional taxonomist': Stephen Jay Gould, 'No science without fancy, no art without facts: The lepidoptery of Vladimir Nabokov', in Funke (1999), p. 86.

p. 129.   'Chinese rhubarb whose root bears an uncanny resemblance': Nabokov (2001), pp. 116–17.

p. 129.   'This passage has been expertly decoded': Dieter E. Zimmer, 'Chinese rhubarb and caterpillars', www.nabokovmuseum.org/PDF/ZimmerR&C.pdf (accessed on 3 December 2008).

p. 129.   'The medicines collected here are rhubarb': A. E. Pratt, *To the Snows of Tibet through China*, 1892, p. 188.

p. 129.   'cunning butterfly in the Brazilian forest': Nabokov (2001), p. 105.

p. 130.   'Several times I shot': Bates (1863), Vol. 1, pp. 181–2.

p. 130.   'the fantastic refinement of "protective mimicry"': Nabokov (2001), p. 354.

p. 130.   'Yet, long before the dawn of mankind': ibid., p. 355.

p. 131.   'In nature as it exists today': ibid., p. 358.

p. 131.   'The mysteries of mimicry had a special attraction for me': Nabokov (2000b), p. 98.

p. 132.   'In high art and pure science, detail is everything': Nabokov, quoted in Philip Zaleski, *Harvard Magazine*, July/August 1986, p. 38.

p. 132.   'Nabokov frequently stated': Stephen Jay Gould, 'No science without fancy, no art without facts: The lepidoptery of Vladimir Nabokov', in Funke (1999), p. 100.

p. 132.   'How incandescently, how incestuously': Nabokov (2000a), p. 343.

p. 132.   'The tactile delights of precise delineation': Stephen Jay Gould, 'No science without fancy, no art without facts: The lepidoptery of Vladimir Nabokov', in Funke (1999), p. 114.

p. 133.   'Certain objects and images are endowed with a comparatively high degree of lyrical force': Caillois (2003), p. 69.

p. 133.   'remarkably anthropomorphic appearance': ibid., p. 73.
p. 134.   'Finally, let us not forget': ibid., p. 79.
p. 134.   '*Hymenopus bicornis*': ibid., p. 80.
p. 134.   'Predators are not at all deceived': ibid., p. 96.
p. 134.   'The objective phenomenon is the fascination': ibid., p. 93.
p. 135.   'Morphological mimicry': ibid., p. 96.
p. 135.   'I should like to have wings': Flaubert (1874), p. 296; translated by PF.
p. 135.   'Whatever the artist may say': Caillois (2003), p. 102.
p. 136.   'Max Ernst's 1921 painting *Celebes*': Roland Penrose, 'Max Ernst's Celebes', 52nd Charlton Lecture, Newcastle, 1969. The next two quotations come from this lecture.
p. 137.   'On the boat I found Julian Trevelyan': Penrose (1981b), p. 124.

Chapter 10 Cannibals and Sunshields

p. 138.   'Yet, while there is the closest analogy': Cott (1940), p. xii.
p. 139.   'In 1935, Kerr had resigned': Hugh B. Cott, 'Sir John Graham Kerr', *Nature*, 1957, 179, pp. 1164–5.
p. 139.   'But Kerr was not completely without influence', *The Times*, 8 June 1939.
p. 139.   'Hugh Bamford Cott (1900–87) was a son of the manse': obituary, *The Times*, 25 April 1987.
p. 140.   'And it is in the tropics': Hugh B. Cott, 'Natural history of the lower Amazon', *Proceedings of the Bristol Natural History Society*, 1930, 4S, 7(3), pp. 181–8.
p. 140.   'In 1933, McAtee unwisely took issue': W. L. McAtee, 'Warning colors and mimicry', *Quarterly Review of Biology*, 1933, 8 (no. 3), pp. 209–13.
p. 140.   'Poisonous secretions are of common occurrence': Hugh B. Cott, 'The Zoological Society's expedition to the Zambesi, 1927: No 4. On the ecology of tree-frogs in the lower Zambezi valley, with special reference to predator habits considered in relation to the theory of warning colours and mimicry', *Proceedings of the Zoological Society of London*, 1932, p. 478.
p. 140.   'There is no evidence of any frog being dangerously poisonous': W. L. McAtee, 'Warning colors and mimicry', *Quarterly Reviews of Biology*, 1933, 8 (no. 3), p. 211.
p. 141.   'Dr McAtee is apparently either unable or unwilling': Hugh B. Cott, 'Warning colours and mimicry', *Proceedings of the Entomological Society of London*, 1935, pp. 109–19.
p. 141–2. 'I am now regarded as an authority': letter from Cott to Kerr, 26 March 1938: JGKA,/DC6/708.
p. 142.   'Cott was not wrong': letter from Cott to Kerr, 16 May 1939: JGKA/DC6/709.
p. 142.   'The confrontation began with a long article by Cott': Hugh B. Cott, 'Camouflage: Nature's hints to man', *The Times*, 30 March 1939.
p. 142.   'Since Kerr maintains he was responsible': *The Times*, 15 April 1939.
p. 142.   'Cott tried to defend': *The Times*, 3 May 1939.
p. 143.   'The link between the film industry': NA/AIR 2/2878.
p. 143.   'the Industrial Camouflage Research Unit': Roland Penrose Archive, Scottish National Gallery of Modern Art, GMA A35/1/1/RPA758.
p. 143 fn. Ernö Goldfinger': Warburton (2004).
p. 144.   'Already the rash of squiggly green patterns': Trevelyan (1957), p. 112.
p. 144.   'a scheme that was in effect': Trevelyan (1957), p. 113.
p. 144.   'the biggest con I was ever involved in': Warburton (2004), p. 99.
p. 144.   'At Mildenhall the hangars': letter from Cott to Kerr, 16 April 1939: JGKA/DC6/709.
p. 145.   'Col Turner (Director of "works and bricks")'; 'these air ministry officials': letter from Cott to Kerr, 9 June 1939: JGKA/DC6/712.
p. 145.   'The party was impressed': letter from Cott to Kerr, 5 July 1939: JGKA/DC6/714.
p. 145.   'Anderson invited Cott': letter from Cott to Kerr, 28 July 1939: JGKA/DC6/715.
p. 145.   'Air Vice Marshal Welsh': letter from Cott to Kerr, 23 August 1939: JGKA/DC6/719.
p. 146.   'A grumpy note': letter from Cott to Kerr, 13 September 1939: JGKA/DC6/720.

p. 146. 'In mid-October the committee': letter from Cott to Kerr, 21 October 1939: JGKA/DC6/724.

p. 146. 'But then in January': letter from Cott to Kerr, 14 January 1940: JGKA/DC6/726.

p. 146. 'Colonel Turner turned his attention': NA/AIR 2/3447.

p. 147. 'With care and attention': NA/AIR 2/3705.

p. 147. 'the 100 night-dummy landing strips': Cruickshank (1979), pp. 5–7.

p. 148. 'In March 1940 Wilkinson fobbed the committee off': undated letter from Cott to Kerr: JGKA/DC6/728.

p. 148. 'he wrote to Anderson': copy letters from Cott to Sir John Anderson, Home Office, 5 April 1940: JGKA/DC6/738; 27 March 1940: JGKA/DC6/733.

p. 148. 'Cott's bid for attention': Hugh B. Cott, 'Camouflage in modern warfare', Nature, 1940, 145 (22 June), pp. 949–51.

p. 149. 'The article drew letters': Nature, 1940, 146 (3 August), pp. 168 and 429.

p. 149. 'Cott had the last word': Nature, 1940, 146 (28 September), p. 429.

p. 149. 'One gun was camouflaged': letter from Cott to Kerr, 19 August 1940: JGKA/DC6/753.

p. 150. 'On 26 October 1940'; 'in a favourable light or angle'; 'In some respects I should prefer': letter from Cott to Kerr, 26 October 1940: JGKA/DC6/758.

p. 151. 'Not everyone at Farnham': Trevelyan (1957), p. 118.

p. 151. 'For six weeks I had to attend lectures': Maskelyne (1949), p. 17.

p. 151. 'Penrose had a Quaker background': Antony Penrose, interview, Farley Farm, East Sussex, 23 May 2008.

p. 151. 'His most appreciated trick': Penrose (1981b), p. 130.

p. 151. 'An excellent short guide': Penrose (1941).

p. 152. 'What does the average city dweller know': Barkas (1952), p. 35.

p. 152. 'Farnham Lecture No. 5': IWM/Major D. A. J. Pavitt/86/50/3.

p. 153. 'Various recent attempts': Cott (1940), p. 53.

p. 154. 'The fact is that in the primeval struggle': ibid., p. 158.

p. 154. 'Observations on the habits': Lecture delivered at SME, Chatham, 20 Oct. 1938: Royal Engineers' Journal, 1938, 52, December, pp. 501–17.

p. 155. 'units arrived with camouflage nets': Sykes (1990), p. 28.

p. 155. 'You cannot hide anything in the desert': Trevelyan (1957), p. 152.

p. 155. 'The resemblance is literally superficial': Cott (1940), pp. 324, 405.

p. 155. 'Cott gives a homely human analogy': ibid., p. 407.

p. 156. 'Barkas entered camouflage work': Barkas (1952), pp. 4–27.

p. 156. 'Barkas established a desert Camouflage Development and Training Camp': 'Middle East Camouflage Report No. 1': NA/ WO201/2843.

p. 156. 'At the school': Trevelyan (1957), p. 154.

p. 156. 'Sergeant Bob Thwaites was a typical NCO': The papers of Sergeant Bob Thwaites: IWM/05/46/1.

p. 157. 'Less enchanted by Helwan': Graham (1974), p. 141. The next three quotations come from this source.

p. 158. 'Though (of course) no one is in the least interested': ibid., p. 143.

p. 158. 'He wrote a searing account': Douglas (2008).

p. 158. 'One member of the group': Maskelyne (1949); Fisher (1985).

p. 159. 'So we got to work again': Maskelyne (1949), p. 121.

p. 159. 'Maskelyne's lurid account': Fisher (1985), p. 369.

p. 159. 'An official Camouflage Report': 'Middle East Camouflage Report No. 1', NA/WO 201/2843.

p. 159. 'He had become locked': undated copy letter from Kerr to Atlee: JGKA/DC6/360.

p. 159. 'A series of verses and cartoons': Fortnightly Fluer, 12 July 1941, in Sykes (1990), pp. 78–9.

p. 160. 'In October 1941, during a week's leave': Cott (1975), p. 185.

p. 161. 'Back on duty': collection of Cott's photographs illustrating camouflage: JGKA/DC6/707.

p. 161. 'a large operation was mounted': Sykes (1990), pp. 41–53; 'Scheme Bertram', NA/WO 201/2023.

p. 162. 'I think that camouflage men': Barkas (1952), p. 146.

p. 162. 'The dummy railhead looks very spectacular': Trevelyan (1957), p. 158.

p. 162. 'was most successful': 'Report on operational camouflage in the Western Desert, August–December 1942': IWM/Captain G. M. Leet 91/2/1; NA/WO 201/2024.

p. 162. 'two-faced locomotive': Sykes (1990), p. 82.

p. 163. 'Another leaf from nature's book': Barkas (1952), pp. 125–6.

p. 163. 'Consider the imitation of oozing poison': Nabokov (2000), p. 98.

p. 163. 'His great offensive': The papers of Viscount Montgomery of Alamein, IWM/BLM27.

p. 163. 'Colonel Dudley Clarke (1899–1974) visited Montgomery's headquarters': 'A' Force Permanent Record File, NA/CAB154/2.

p. 163. 'he led a charmed life': Rankin (2009), pp. 345–8.

p. 164. 'Ayrton and I trudged': Barkas (1952), p. 194.

p. 164. 'The brief for Operation Bertram': 'Scheme Bertram', NA/WO 201/2023.

p. 165. 'The Sunshields were the most successful': NA/WO201/2841.

p. 166. 'Many animals habitually masquerade': Cott (1940), p. 358.

p. 167. 'Such tactics offer wide scope': Ibid., p. 359.

p. 167. 'In nature many animals': Farnham Camouflage Notes 1943, IWM/Maj. D. A. J. Pavitt, 86/50/3.

p. 168. 'It was a wonderful sight': The papers of Viscount Montgomery of Alamein, IWM/BLM27.

p. 169. 'By a marvellous system of camouflage': *Hansard*, HC Deb 11 November 1942, Vol. 385, cc37. http://hansard.millbanksystems.com/commons/1942/nov/11/debate-on-the-address (accessed 11 March 2009).

p. 169. 'Though none of us was so foolish': Barkas (1952), p. 216.

Chapter 11   Dazzle (Revisited) to D-Day

p. 170. 'As a boy of twelve': Sir Scott, letter to Mrs Mary F. Boynton, 10 September 1950, in White (1951), pp. 137–8.

p. 172. 'Peter Scott was the son': Huxley (1993).

p. 172. 'In July 1940 Scott had his own ship': PS/B15.

p. 172. 'Compromise is usually fatal in a camouflage scheme': PS/B17.

p. 172. 'He stressed that the aim': PS/B15.

p. 173. 'Camouflage of destroyers': 'Memo to all ships: Camouflage of vessel operating against U boats': 16 May 1941, PS/B17.

p. 173. 'It is possible that, where designs': 'Naval Camouflage', NA/HO 217/9.

p. 174. 'This work convinced the Admiralty': *The Camouflage of Ships at Sea*, C.B. 3098, Training and Staff Duties Division, Naval Staff, Admiralty, May 1943.

p. 174. 'In 1942 he was still pleading with the Admiralty': letter from Kerr to Sir Victor Warrender, Admiralty, 13 January 1942: JGKA/DC6/584.

p. 174. 'In the US Navy': Williams (2001), pp. 136–203.

p. 174. 'He was one of the earliest advocates': White (1951), pp. 138–9.

p. 175. 'Towards the end of the war': 'Trials of British Admiralty and US Navy camouflage measures, Chesapeake Bay, October 1944', NA/ADM 212/137.

p. 176. 'There can in fact be no such thing'; 'Each type achieves its primary objective'; 'administrative difficulties': ibid.

p. 176. 'A fascinating piece of research': 'Countershading: Report on tests made on a new method', NA/ADM 212/135.

p. 177. 'Under a plan code-named "Fortitude" ': NA/WO 219/2233.

pp. 177–8. 'The plan calls for protective decoys'; 'Menabilly was sited': NA/AIR 2/6022.

p. 178. 'The dummy landing craft': NA/WO 199/2629.

p. 179. 'Other deceptions': Rankin (2009), pp. 399–402, 407.

p. 179. 'On the day itself': NA/AIR 14/2041.

p. 179. 'But the larger deception': NA/WO 219/2233; Rankin (2009), pp. 407–8.

p. 180. ' Mass production of camouflaged uniforms': Newark (1996), pp. 21–6.

p. 181. 'On that day': Miller (1992), p. 73.

p. 181. 'Picasso would have imagined Roland'; 'They're all the same thing': Antony Penrose, interview, Farley Farm, East Sussex, 23 May 2008.

p. 181. 'God is really only another artist': Gilot (1966), p. 43.

Chapter 12   From Butterflies to Babies and Back

p. 182. 'As the laws of nature': Bates (1863), Vol. 2, p. 346.

p. 182. 'Since 1946, Philip Sheppard': *Biographical Memoirs of the Royal Society*, 1977, 23, pp. 465–500.

p. 183. 'One of his adversaries, J. B. S. Haldane': EBF/D4.

p. 184. 'I very much hope': EBF/F5.

p. 184. 'A polymorphism was succinctly defined by Ford': Ford (1955), p. 63.

p. 185. 'In 1936 Ford had published': E. B. Ford, 'The Genetics of *Papilio dardanus*', *Transactions of the Entomological Society of London*, 1936, 85, pp. 435–66.

p. 185. 'After six years in the Navy in Australia': Sir David Weatherall, 'Sir Cyril Astley Clarke, CBE,' *Biographical Memoirs of the Royal Society*, 2002, 48, pp. 69–85.

p. 185. 'in an idle moment one Sunday afternoon': C. A. Clarke, 'The Prevention of, "Rhesus" Babies,' *Scientific American*, 1968, 219, p. 46.

p. 186. 'In 1953 the experiment produced': C. A. Clarke, 'A hybrid swallowtail'. *Entomologist's Journal of Record and Variation*, 1953, 65, pp. 76–80.

p. 186. 'we met for the first time': Sir Cyril Clarke, 'Philip Macdonald Sheppard', *Biographical Memoirs of Fellows of the Royal Society*, 1977, 23, p. 478.

p. 186. 'wasn't tough enough to vault': ibid., p. 468.

p. 186. 'Blood groups': ibid., p. 486.

p. 186. 'We could not help noticing': C. A. Clarke, 'Prevention of Rh-haemolytic disease', *British Medical Journal*, 1967, 4 (no. 5570), pp. 7–12.

p. 187. 'As Clarke tells the story': C. A. Clarke, 'The Prevention of Rhesus Babies', *Scientific American*, 1968, 219, p. 49.

p. 188. 'The circumstances of the first treatment': *The Times*, 4 March 1964.

p. 188. 'In 1965–6 the results': Third Annual Report, Nuffield Unit of Medical Genetics. EBF/F15.

p. 188. 'The results can be seen': L. A. Derrick Tovey, 'The contribution of antenatal anti-D prophylaxis to the reduction of the morbidity and mortality in Rh haemolytic disease of the newborn', *Plasma Therapy and Transfusion Technology*, 1984, 5, pp. 99–100.

p. 189. 'When I discovered that story': Michael Majerus, interview, Cambridge, 27 November 2007.

p. 189. 'The new building': letter from Clarke to Ford, 1 March 1967: EBF/F13.

p. 189. 'An amusing correspondence': for quotations in the next two paragraphs, see EBF/F13.

p. 190. 'It is necessary to suppose': Darwin (1951), p. 493.

p. 191. 'The odds for the butterfly': E. B. Ford, 'Genetics of polymorphism in lepidoptera', *Advances in Genetics*, 1953, 5, p. 63.

p. 191. 'Also in 1953, Ford': ibid., pp. 70–1.

p. 192. 'Goldschmidt had rashly and brashly predicted': Richard Goldschmidt, 'Mimetic polymorphism, a controversial chapter of Darwinism', *Quarterly Review of Biology*, 1945, 20, p. 213.

p. 192. 'As far as colour and pattern': C. A. Clarke and P. M. Sheppard, 'Super-genes and mimicry', *Heredity*, 1960, 14, p. 175.

p. 193. 'This mode of genetic determination': ibid.

p. 193. 'The best evidence for this came from Madagascar': C. A. Clarke and P. M. Sheppard, 'Interactions between major genes and polygenes in the determination of the mimetic pattern of *Papilio dardanus*', *Evolution*, 1963, 17, pp. 404–13.

p. 194. 'But in 1975 a husband-and-wife team': D. Charlesworth and B. Charlesworth, 'Theoretical genetics of Batesian mimicry. II. Evolution of supergenes', *Journal of Theoretical Biology*, 1975, 55, pp. 305–24.

p. 195. 'Many people will find it improbable': John R. G. Turner, 'Mimicry: The palatability spectrum and its consequences', in Vane-Wright and Ackery (eds), (1984), pp. 141–61.

p. 196. 'the mammoth *Heliconius* paper': P. M. Sheppard, J. R. G. Turner et al., 'Genetics and the evolution of muellerian mimicry in *Heliconius* butterflies': *Philosophical Transactions of the Royal Society of London*, 1985, 308, pp. 433–613.

Chapter 13 The Aromas of Mimicry

p. 197. 'plenishing, domestic smells': Proust (1972), p. 65.

p. 197. 'Officially she was not a member': 'Dame Miriam Louisa Rothschild. 5 August 1908–20 January 2005', *Biographical Memoirs of the Royal Society*, 2006, 52, pp. 315–30.

p. 198. 'When he came to lunch one day': Miriam Rothschild, 'Speculations about mimicry with Henry Ford', in Creed (ed.), (1971), p. 202.

p. 198. 'Squeeze a ladybird': *Guardian*, obituary of Miriam Rothschild, 22 January 2005.

p. 199. 'One of the curious facets': Miriam Rothschild, 'Secondary plant substances and warning colouration in insects', in van Emden (ed.), (1973), p. 70.

p. 199. 'a virtuoso natural products chemist': Miriam Rothschild, 'Tadeus Reichstein, 20 July 1897–1 August 1996', *Biographical Memoirs of the Royal Society*, 1999, 45, pp. 451–67.

p. 200. 'This was the first time that a digitalis-like toxin': T. Reichstein, J. von Euw, J. A. Parsons and Miriam Rothschild, 'Heart poisons in the monarch butterfly', *Science*, 1968, 161 (no. 3844), pp. 861–66.

p. 200. 'Rothschild and Reichstein gave a presentation': M. Rothschild et al., 'Poisons in aposematic insects', exhibit no. 19, Conversazione, The Royal Society, London, 1966.

p. 201. 'Apart from the unpalatable nature': Trimen (1887), p. 54.

p. 201. 'There seems little doubt': Miriam Rothschild, 'Some observations on the relationship between plants, toxic insects and birds', in Harborne (ed.), (1972), p. 3.

p. 201. 'who introduced the concept': Eisner (2003), p. 22.

p. 202. 'Eisner discovered the bombardier': ibid., pp. 9–43.

p. 203. 'Poison-spraying beetles': R. B. Huey and E. R. Pianka, 'Natural selection for juvenile Lizards Mimicking Noxious Beetles', *Science*, 1977, 195, pp. 201–3.

p. 204. 'Rothschild had the answer': Miriam Rothschild, 'Aide Memoire Mimicry', *Ecological Entomology*, 1984, 9, pp. 311–19.

p. 204. 'The colour of day flying Lepidoptera': letter from Rothschild to Ford, 2 September 1974: EBF/D16.

p. 205. 'In a letter to his acolyte': letter from Ford to Sheppard, 21 June 1968: EBF.

p. 205. 'In very large part this', R. C. Lewontin, 'Testing the theory of natural selection', *Nature*, 1972, 236, p. 181.

p. 206. 'silly toffs with butterfly nets': Hooper (2002), p. 216.

p. 206. 'There is no doubt': Haldane, review of *Ecological Genetics*: EBF/D16.

p. 206. 'There have been great successes': R. C. Lewontin, 'Testing the theory of natural selection', *Nature*, 1972, 236, p. 182.

Chapter 14 The Tinkerer's Palette

p. 207. 'God is just another artist': Françoise Gilot, *Life with Picasso*, 1966, p. 43.

p. 207. 'he didn't give me a name': Richard Feynman, Radio Broadcast, ABC, 26 June 2008; www.abc.net.au/rn/inconversation/stories/2008/2276846.htm (accessed 13 December 2008).

p. 209. 'The science of pattern formation': Carroll (2005).

p. 211. 'What is it that kicks the dominos off?': Sean Carroll, interview, London, 2 April 2008.

p. 211. 'The spirit is too blunt an instrument': Stevenson (2005), p. 122.

p. 212.   'Armageddon pierstaff!': Amis (1968).

p. 212.   'In 2007 Craig Venter': 'The diploid sequence of an individual human', *PloS Biology*, 2007, 5(10):e254 doi:10.1371/journal.pbio.0050254.

p. 212 fn. 'Mobile genetic elements': McClintock (1987); Dover (2001), pp. 31, 32.

p. 213.   'The hypothesis that explains mimicry': E. B. Poulton, *Bedrock*, 1913, No. 1, p. 65.

p. 215.   'Carroll's team was looking at the expression': Sean B. Carroll et al., 'Pattern formation and eyespot determination in butterfly wings', *Science*, 1994, 265, pp. 109–14.

p. 215.   'Carroll's team also showed that the basic unit': Ron Galant et al., 'Expression pattern of a butterfly achaete-scute homolog reveals the homology of butterfly wing scales and insect sensory bristles', *Current Biology*, 1998, 8 (no. 14), pp. 807–13.

p. 216.   'Carroll addressed the coloration of a species of fruit fly': Sean B. Carroll, et al., 'Regulating evolution', *Scientific American*, 2008, 298 (May), pp. 38–45.

p. 216.   'Carroll has discovered that, if you disable a regulatory gene': Sean Carroll, interview, London, 2 April 2008.

p. 217.   '*Svb* [the shaven baby gene] seems peculiar': Alistair P. McGregor et al., 'Morphological evolution through multiple *cis*-regulatory mutations at a single gene', *Nature*, 2007, 448, pp. 587–90.

p. 217.   'Our results provide experimental evidence': ibid., p. 589.

p. 218.   'the butterfly research community was divided': Antonia Monteiro, personal communication, 23 April 2008.

p. 218.   '454 sequencing': Wikipedia: http//en.wikipedia.org/wiki/454_Life_Sciences (accessed 13 December 2008).

p. 219.   'The stickleback uses armour': P. F. Colosimo et al. 'Widespread parallel evolution in sticklebacks by repeated fixation of ectodysplasin alleles', *Science*, 2005, 307, pp. 1928–33.

p. 220.   'Our results show': ibid., p. 1933.

Chapter 15   The *Heliconius* Variations

p. 221.   'Such colours can't be drawn': Joseph Brodsky, 'The Butterfly' (1980), pp. 68–9.

p. 221.   '*Heliconius* butterflies are Müllerian mimics': P. M. Sheppard, J. R. G. Turner et al., 'Genetics and the evolution of muellerian mimicry in *Heliconius* butterflies', *Philosophical Transactions of the Royal Society B*, 1985, 308, pp. 433–613.

p. 221.   'I think it likely': Hardy (1965), p. 151.

p. 222.   '*Numata* is highly polymorphic': K. S. Brown and W. W. Benson, 'Adaptive polymorphism associated with multiple Müllerian mimicry in *Heliconius numata*', *Biotropica*, 1974, 6, pp. 205–28.

p. 222.   'Dennis the Menace': *Philosophical Transactions of the Royal Society B*, 1985, 308, p. 457.

p. 225.   'In 2006 the researchers found that the colour-pattern genes': Mathieu Joron et al., 'A conserved supergene locus controls colour pattern diversity in *Heliconius* butterflies', *PloS Biology*, 2006, 4 (no. 10), e303 doi:10.1371/journal.pbio.0040303.

p. 225.   'The favourite is a kinesin gene': Chris Jiggins, personal communication, 1 January 2009.

p. 226.   'the red-eye pigment in fruit flies': R. D. Reed and L. M. Nagy, 'Evolutionary redeployment of a biosynthetic module: Expression of eye pigment genes vermilion, cinnabar, and white in butterfly wing development', *Evolution and Development*, 2005, 7, pp. 301–11.

p. 226.   'formation of the red scales': Robert D. Reed et al., 'Gene expression underlying adaptive radiation on *Heliconius* wing patterns: Non-modular regulation of overlapping cinnabar and vermilion prepatterns', *Proceedings of the Royal Society B*, 2008, 275, pp. 37–45.

p. 227.   'the Jiggins team then set to work': Laura D. Ferguson and Chris D. Jiggins, 'Both shared and divergent expression domains underlie a convergent mimetic phenotype in *Heliconius* butterflies', to be published in *Evolution and Development*.

p. 228.   'The basic biology of *cydno* and *melpomene* is very similar': James Mallet, 'Rapid speciation, hybridization, and adaptive radiation in the *Heliconius melpomene* group', in Butlin, Bridle and Schluter (eds), (2009), pp. 177–94.

p. 228. 'Convergence of colour patterns/schemes': letter from Miriam Rothschild to E. B. Ford, 2 September 1974: EBF/D16.

p. 229. 'In the 1990s some strange findings began to emerge': A. V. Z. Brower, 'A new mimetic species of Heliconius (Lepidoptera: Nymphalidae), from south-eastern Colombia, revealed by cladistic analysis of mitochondrial DNA sequences', *Zoological Journal of the Linnean Society*, 1996, 116, pp. 317–32.

p. 229. 'Mauricio followed the trail': Mauricio Linares, interview, London, 25 June 2007.

p. 230. 'This slowed the chase down somewhat': James Mallet, 'Rapid speciation, hybridization and adaptive radiation in the *Heliconius melpomene* group', in Butlin, Bridle and Schluter (eds), (2009), pp. 187–9.

p. 230. 'a 1998 expedition to Churumbelos': 'Colombia' 98 Expedition to Serrania de los Churumbelos,http://www.proaves.org/IMG/pdf/EBA_1_Churumbelos_report_1998. pdf (accessed 12 December 2008); Blanca Huertas, interview, Natural History Museum, London, 26 September 2008.

p. 230. 'The results were published in 2008': Nathalia Giraldo et al., 'Two sisters in the same dress: *Heliconius* cryptic species', *BMC Evolutionary Biology*, 2008, 8, p. 324; published online 28 November 2008 at doi: 10.1186/1471-2148-8-324.

p. 230. 'achieving newspaper headlines in 2006': 'Butterfly effect: New species hatches in lab', *Guardian*, 15 June 2006.

p. 231. 'It seems likely that isolation as species': C. D. Jiggins et al., 'Reproductive isolation caused by colour pattern mimicry', *Nature*, 2001, 411, pp. 302–5.

p. 231. 'a strong predilection to pair together', Darwin (1977), p. 92.

p. 231. 'The lab-bred *H. heurippa*': J. Mavarez et al., 'Speciation by hybridization in *Heliconius* butterflies', *Nature*, 2006, 441, pp. 868–71; Mauricio Linares, 'Homoploid hybrid evolution in *Heliconius* butterflies', paper read at symposium 'The Driving Forces of Evolution', Linnean Society, 3 July 2008; James Mallet, 'Rapid speciation, hybridization, and adaptive radiation in the *Heliconius melpomene* group', in Butlin, Bridle and Schluter (eds), (2009), pp. 177–94.

p. 231. 'these "artificial" butterflies preferred to mate with their own kind': Maria C. Melo et al., 'Assortative mating preferences among hybrids offers a route to hybrid speciation', *Evolution*, in press.

p. 232. 'Of present-day *Heliconius* species, 20 per cent': Lawrence E. Gilbert, 'Adaptive novelty through introgression in *Heliconius* wing patterns', in Boggs, Watt and Ehrlich (eds), (2003), pp. 281–318.

p. 233. 'The males display a sexual ardour': Miriam Rothschild, *Butterfly Cooing Like Dove*, 1991, p. 180.

p. 233. 'the real joker in the pack is *Heliconius numata*': Mathieu Joron et al., 'A conserved supergene locus controls colour pattern diversity in *Heliconius* butterflies', *PloS Biology*, 2006, 4 (no. 10), e303 doi:10.1371/journal.pbio.0040303.

Chapter 16 A Shifting Spectrum

p. 235. 'These animals also escape detection': Darwin (1989), p. 46.

p. 236. 'chameleon colours have some functions': Kira O'Day, 'Conspicuous chameleons', *PloS Biology*, 2008, 6(1): e21 doi:10.1371/journal.pbio.0060021; D. Stuart-Fox, and A. Moussalli, 'Camouflage and colour change: Antipredator responses to bird and snake predators across multiple populations in a dwarf chameleon', *Biological Journal of the Linnean Society*, 2006, 88, pp. 437–45.

p. 237. 'Why *these* animals, and how do they do it?': Roger Hanlon, 'Cephalopod dynamic camouflage', *Current Biology*, 2007, 17 (no. 11), R400–4.

p. 238. 'The new knowledge about octopuses': Roger Hanlon et al., 'Mimicry and foraging behaviour of two sand-flat octopus species off North Sulawesi, Indonesia', *Biological Journal of the Linnean Society*, 2008, 93 (no. 1), pp. 23–38.

p. 239.   'A good test case involves the crab spiders': Astrid M. Heiling, Marie E. Herberstein and Lars Chittka, 'Pollinator attraction: Crab spiders manipulate flowers signals', *Nature*, 2003, 421, p. 334.

p. 240.   'But the strangest of Herberstein's discoveries': A. M. Heiling and M. E. Herberstein, 'Predator–prey coevolution: Australian native bees avoid their spider predators', *Proceedings of the Royal Society B* (Suppl.), 2004, 271, S196–S198.

p. 241.   'One of the most sinuous cases': David W. Pfennig, 'Frequency dependent Batesian mimicry', *Nature*, 2001, 410, p. 323; George R. Harper Jr and David W. Pfennig, 'Mimicry on the edge: Why do mimics vary in resemblance to their model in different parts of their geographical range?' *Proceedings of the Royal Society B*, 2007, 274, pp. 19655–61.

p. 242.   'There may not have been strong selection': personal communication, 20 March 2008.

p. 242.   'the case of industrial pollution and the peppered moth': Hooper (2002).

p. 244.   'Following Kettlewell': J. S. Jones, 'More to melanism than meets the eye', *Nature*, 1982, 300, pp. 109–10.

p. 245.   'In that year, the late Michael Majerus': Majerus (1998).

p. 245.   'When *Melanism* was reviewed in *Nature*': 'Not black and white,' review of Michael E. N. Majerus, *Melanism: Evolution in Action, Nature*, 1998, 396, pp. 35–6.

p. 246.   'My view of the rise and fall': Majerus (1998), p. 155.

p. 246.   'Evolution experts are quietly admitting': *Sunday Telegraph*, 14 March 1999.

p. 247.   'stunted social skills': Hooper (2002), p. 9.

p. 247.   'Majerus ran an experiment': E. N. Majerus, 'The peppered moth: The proof of Darwinian evolution', www.gen.cam.ac.uk/Research/Majerus/Swedentalk220807.pdf (accessed 2 March 2009).

p. 247.   'He found just over 50 per cent': E. N. Majerus, 'The peppered moth: The proof of Darwinian evolution, www.gen.cam.ac.uk/Research/Majerus/Swedentalk220807.pdf (accessed 2 March 2009).

p. 247 fn.   'Some moths have evolved countermeasures': Nickolay I. Hristov and William E. Connor, 'Sound strategy: Acoustic aposematism in the bat-tiger moth arms race', *Naturwissenschaften*, 2005, 92, pp. 164–9.

p. 248.   'He marked, released and recaptured the moths': ibid.

p. 248.   'Jerry Coyne told Majerus': Michael Majerus, interview, Cambridge, 27 November 2007.

Epilogue

p. 251.   'the rock pocket mice': Michael W. Nachman, Hopi E. Hoekstra Susan L. D'Agostino, 'The genetic basis of adaptive melanism in pocket mice', *Proceedings of the National Academy of Sciences*, 2003, 100, pp. 5, 268–73.

p. 252.   'But in the 1960s the Vietnam War': 'Camouflage US Marine Corps utility uniform: Pattern fabric and design, 2004', US Patent 6805957.

p. 255.   'In the last few years, there has been an explosion': Martin Stevens and Sami Merilaita, 'Animal camouflage: Current issues and new perspectives', *Philosophical Transactions of the Royal Society B*, 2009, 364, pp. 423–7.

BIBLIOGRAPHY

Addison, Colonel G. H. (1926). *The Work of the Royal Engineers in the European War, 1914–1918*, Miscellaneous, Institution of Royal Engineers, Chatham.

Allen, Garland E., (1978). *Thomas Hunt Morgan*, Princeton University Press, Princeton.

Amis, Kingsley (1968). *I Want it Now*, Jonathan Cape, London.

Anderson, Ross (1982). *Abbott Handerson Thayer*, Everson Museum, Syracuse, New York.

Barkas, Geoffrey (1952). *The Camouflage Story*, Cassell Plc, a division of the Orion Publishing Group.

Bates, Henry Walter (1863). *The Naturalist on the River Amazons*, Vols 1 and 2, John Murray, London.

Bates, Henry Walter (1892). *The Naturalist on the River Amazons*, John Murray, London.

Bateson, William (1894). *Materials for the Study of Variation*, Macmillan, London.

Beddard, Frank E. (1892). *Animal Coloration*, Swan Sonnenschein, London.

Belt, Thomas (1874). *The Naturalist in Nicaragua*, John Murray, London.

Blunt, Wilfred Scawen (1919–20). *My Diaries: Being a Personal Narrative of Events, 1888–1914*, Vol. 2, Martin Secker, London.

Boggs, Carol. L., Watt, Ward B. and Ehrlich, Paul R. (eds) (2003). *Butterflies: Ecology and Evolution Taking Flight*, University of Chicago Press, Chicago.

Brodsky, Joseph (1980). *A Part of Speech*, Oxford University Press, Oxford; Farrar, Strauss and Giroux, New York.

Brown, Milton (1963). *The Story of the Armory Show*, Joseph H. Hirshhorn Foundation, Greenwich, CT.

Browne, Janet (2002). *Charles Darwin: The Power of Place*, Jonathan Cape, London.

Burchell, William J. (1822). *Travels in the Interior of Southern Africa*, Vol. I, Longmans, London.

Butlin, Roger K., Bridle, Jon and Schluter, Dolph (eds) (2009). *Speciation and Patterns of Diversity*, Cambridge University Press, Cambridge.

Caillois, R. (2003). *The Edge of Surrealism: A Roger Caillois Reader*, ed. Claudine Frank, Duke University Press, Durham, NC.

Carlson, Elof Axel (2004). *Mendel's Legacy*, Cold Spring Harbor Laboratory Press, New York.

Carroll, Sean B. (2005). *Endless Forms Most Beautiful*, W. W. Norton, New York.

Carroll, Sean B. (2008). *The Making of the Fittest*, Quercus, London.

Cortissoz, Royal (1923). *American Artists*, Charles Scribner's, New York.

Cott, Hugh B. (1940). *Adaptive Coloration in Animals*, Methuen, London.

Cott, Hugh B. (1975). *Looking at Animals*, Collins, London.

Creed, Robert (ed.) (1971). *Ecological Genetics and Evolution*, Blackwell Scientific Publications, Oxford.

Cruickshank, Charles (1979). *Deception in World War II*, Oxford University Press, Oxford.

Darwin, Charles (1862). *On the Various Contrivances by which British and Foreign Orchids are fertilised by Insects*, John Murray, London.

Darwin, Charles (1868). *The Variation of Animals and Plants under Domestication*, John Murray, London.

Darwin, Charles (1871). *The Descent of Man*, Vol. I, John Murray, London.

Darwin, Charles (1951). *The Origin of Species*, a reprint of 6th edition, Oxford University Press, Oxford.

Darwin, Charles (1977). *Collected Papers*, ed. Paul H. Barrett, University of Chicago Press, Chicago.

Darwin, Charles (1985). *The Origin of Species*, Penguin, London.

Darwin, Charles (1989). *The Voyage of the Beagle*, Penguin, London.

Darwin, Charles (2002). *Autobiographies*, Penguin, London.

Darwin, Erasmus (1796). *Zoonomia; or the Laws of Organic Life*, Vol. 1, J. Johnson, London.

Dawkins, Richard (1988). *The Blind Watchmaker*, Penguin, London.

Desmond, Adrian and Moore, James (1992). *Darwin*, Penguin, Harmondsworth.

de Vries, Hugo (1909). *The Mutation Theory*, trans. J. B. Farmer and A. D. Darbishire, Vol. 1, Open Court Publishing, Chicago.

Douglas, Keith (2008). *Alamein to Zem Zem*, Faber and Faber, London.

Dover, Gabriel (2001). *Dear Mr Darwin*, Phoenix, London.

Edwards, William H. (1847). *A Voyage up the Amazon*, John Murray, London.

Eisner, Thomas (2003). *For Love of Insects*, Belknap Press of Harvard University Press, Cambridge, MA.

Faulkner, Barry (1973). *Sketches from an Artist's Life*, William L. Bauhan, Dublin, NH.

Fisher, David (1985). *The War Magician*, Corgi, London.

Fisher, R. A. (1930). *The Genetical Theory of Natural Selection*, Clarendon Press, Oxford.

Flaubert, Gustave (1874). *La Tentation de Saint Antoine*, Charpentier et Cie, Paris.

Forbes, Henry O. (1989). *A Naturalist's Wanderings in the Eastern Archipelago* (facsimile of 1885 edition), Oxford University Press, Oxford.

Ford, E. B. (1955). *Moths*, Collins, London.

Ford, E. B. (1964). *Ecological Genetics*, Methuen, London.

Ford, E. B. (1967). *Butterflies*, Collins, London.

Funke, Sarah (ed.) (1999). *Vera's Butterflies*, Glen Horowitz Bookseller, New York.

Gilot, Françoise (1966). *Life with Picasso*, Penguin, Harmondsworth.

Goldschmidt, Richard (1938). *Physiological Genetics*, McGraw-Hill, New York.

Goldschmidt, Richard (1960). *In and Out of the Ivory Tower*, University of Washington Press, Seattle.

Goldschmidt, Richard (1982). *The Material Basis of Evolution*, Yale University Press, New Haven and London.

Gombrich, E. H. (1982). *The Image and the Eye*, Phaidon Press, London.

Gould, Stephen Jay (1982). *The Panda's Thumb*, W. W. Norton, New York.

Graham, Desmond (1974). *Keith Douglas, 1920–1944: A Biography*, Oxford University Press, Oxford.

Gregory, John Walter (1896). *The Great Rift Valley*, John Murray, London.

Grene, Marjorie (1983). *Dimensions of Darwinism*, Cambridge University Press, Cambridge.

Harborne, J. B. (ed.) (1972). *Phytochemical Ecology: Proceedings of the Phytochemical Society Symposium*, Royal Holloway College, Englefield Green, Surrey, April 1971, Phytochemical Society.

Hardy, Sir Alister (1965). *The Living Stream*, Collins, London.

Harrison, Frederick W. and Locke, Michael (eds) (1999). *Microscopic Anatomy of Invertebrates. Insecta*, Wiley-Liss, New York.

Henig, Robin Marantz (2000). *A Monk and Two Peas*, Weidenfeld & Nicolson, London.

Hooper, Judith (2002). *Of Moths and Men*, Fourth Estate, London.

Huxley, Elspeth (1993). *Peter Scott*, Faber and Faber, London.

Levi, Primo (1985). *The Periodic Table*, Michael Joseph, London (original publication Schocken, New York, 1984).

Liberman, Alexander (1994). *The Artist in His Studio*, Random House Value Publishing, New York (original publication Thames and Hudson, London, 1960).

Lull, Richard Swann (1917). *Organic Evolution*, Macmillan, New York.

Lysenko, T. D. (1954). *Agrobiology*, Foreign Languages Publishing House, Moscow.

Mabey, Richard (1983). *In a Green Shade*, Hutchinson, London.

McClintock, B. (1987). *The Discovery and Characterization of Transposable Elements*, Garland Publishing, New York.

Majerus, Michael (1998). *Melanism*, Oxford University Press, Oxford.

Mare, André (1996). *Carnets de Guerre 1914–1918*, presenté par Laurence Graffan, Herscher, Paris.

Maskelyne, Jasper (1949). *Magic – Top Secret*, Stanley Paul, London.

Miller, Lee (1992). *Lee Miller's War*, ed. Anthony Penrose, Conde Nast, London.

Moon, H. P. (1976). *Henry Walter Bates FRS, 1825–1892*, Leicestershire Museums, Art Galleries and Records Service, Leicester.

Nabokov, Vladimir (2000a). *Ada or Ardor*, Penguin, London.

Nabokov, Vladimir (2000b). *Speak, Memory*, Penguin, London.

Nabokov, Vladimir (2001). *The Gift*, with a new addendum, trans. Dimitri Nabokov, Penguin, London.

Newark, Tim (1996). *Brassey's Book of Camouflage*, Brassey's, London.

Newark, Tim (2007). *Camouflage*, Thames and Hudson, London.

Nijhout, H. Frederick (1991). *The Development and Evolution of Butterfly Wing Patterns*, Smithsonian Institution Press, Washington, DC.

Owen, Richard, *The Zoology of the Voyage of HMS Beagle*, Royal Geographical Society, 1994.

Penrose, Roland (1941). *The Home Guard Manual of Camouflage*, George Routledge, London.

Penrose, Roland (1981a). *Picasso, His Life and Work*, 3rd edition, Granada, London.

Penrose, Roland (1981b). *Scrap Book 1900–1981*, Thames and Hudson, London.

Phillips, Olga Somech (1933). *Solomon J. Solomon*, Herbert Joseph, London.

Poulton, Edward Bagnall (1890). *The Colours of Animals* (the International Scientific Series, Vol. 68), Kegan Paul Trench, Trübner, London.

Poulton, Edward Bagnall (1908). *Essays on Evolution 1889–1907*, Clarendon Press, Oxford.

Proust, Marcel (1972). *Remembrance of Things Past, Swann's Way Pt 1*, trans. C. K. Scott Moncrieff, Chatto & Windus, London.

Punnett, Reginald Crundall (1915). *Mimicry in Butterflies*, Cambridge University Press, Cambridge.

Rankin, Nicholas (2009). *Churchill's Wizards*, Faber and Faber, London.

Roosevelt, Theodore (1910). *African Game Trails*, Charles Scribner's, New York.

Solomon, Solomon J. (1920). *Strategic Camouflage*, John Murray, London.

Stecher, Robert M. (1969). 'The Darwin–Bates Letters – Correspondence between two nineteenth-century travellers and naturalists, Part 1', *Annals of Science*, 25, No. 1, pp. 1–47, 95–125.

Stein, Gertrude (1948). *Picasso*, B. T. Batsford, London.

Stevenson, Anne (2005). *Poems 1955–2005*, Bloodaxe, High Green.

Sutherland, S. K. (1983). *Australian Animal Toxins*, Oxford University Press, Melbourne.

Sykes, Steven (1990). *Deceivers Ever: Memoirs of a Camouflage Officer*, Spellmount, Tunbridge Wells.

Thayer, Gerald H. (1909). *Concealing Coloration in the Animal Kingdom*, Macmillan, New York.

Thayer, Gerald H. (1918). *Concealing Coloration in the Animal Kingdom*, 2nd edition, Macmillan, New York.

Trevelyan, Julian (1957). *Indigo Days*, MacGibbon & Kee, London.

Trimen, Roland (1887). *South African Butterflies*, Vol. 1, Trübner, London.

Van Emden, H. F. (ed.) (1973). *Insect/Plant Relationships* (Proceedings of the 6th Symposium of the Royal Entomological Society of London), Blackwells Scientific, Oxford.

Vane-Wright, R. I. and Ackery, P. R. (eds) (1984). *The Biology of Butterflies* (Royal Entomological Society of London Symposium. No. 11), Academic Press, New York.

Wallace, Alfred Russel (1853). *A Narrative of Travels on the Amazon and Rio Negro*, Reeve, London.

Wallace, Alfred Russel (1869). *The Malay Archipelago*, Macmillan, London.

Wallace, Alfred Russel (1908). *My Life*, Chapman & Hall, London.

Warburton, Nigel (2004). *Ernö Goldfinger: The Life of an Architect*, Routledge, London.

Wells, H. G. (1998). *The Complete Stories of H. G. Wells*, Dent, London.

West, David A. (2003). *Fritz Müller: A Naturalist in Brazil*, Pocahontas Press, Blacksburg, VA.

White, Nelson C. (1951). *Abbott H. Thayer*, Connecticut Printers, CT.

Wilkinson, Norman (1969). *A Brush with Life*, Seeley, London.

Williams, David (2001). *Naval Camouflage 1914–1945: A Complete Visual Reference*, Chatham Publishing, London.

INDEX

WITHDRAWN